REALITY
VS
QUANTUM MYSTICISM

An Attempt to Resolve Issues with Relativity and Quantum Mechanics and Explain Dark Energy

R. CURTIS ARTHUR

 FriesenPress

Suite 300 - 990 Fort St
Victoria, BC, V8V 3K2
Canada

www.friesenpress.com

Copyright © 2021 by R. Curtis Arthur
First Edition — 2021

All rights reserved.

No part of this publication may be reproduced in any form, or by any means, electronic or mechanical, including photocopying, recording, or any information browsing, storage, or retrieval system, without permission in writing from FriesenPress.

ISBN
978-1-5255-7432-0 (Hardcover)
978-1-5255-7433-7 (Paperback)
978-1-5255-7434-4 (eBook)

1. SCIENCE, PHYSICS

Distributed to the trade by The Ingram Book Company

CONTENTS

Author's Preliminary Comments v
Preface vii
Acknowledgments xi

Chapter 1 The Quantum Enigma 1
    1A Introduction 1
    1B The Enigma 3

Chapter 2 Contribution of the Theories of Relativity
to the Quantum Enigma 15
    2A The Limitations of Mathematics 16
    2B Misleading Aspects of the Special Theory of Relativity 18
    2C Misleading "Revolutionary" New Concepts of
Time and Space 24
    2D The Problems with the General Theory of Relativity 31
    2E Further Evidence Against the Curvature of Space
as an Explanation for Gravity 37

Chapter 3 Critical Analysis of the Experimental Evidence for
Relativity and Quantum Mechanics and Related Theories 45
    3A String Theory, Multiple Dimensions and Parallel Worlds 45
    3B The Big Bang Theory and Inflation 48
    3C The Expanding Universe 49
    3D Decoherence 50
    3E Gravitational Lensing as a Test of General Relativity 52
    3F Observation of Gravitational Waves 53
    3G Gravity Probe B 54
    3H Quantum Entanglement and Bell's Theorem
and Inequality 55

Chapter 4	Summary of Findings and Conclusions	59
Chapter 5	A Stab at a New Theory of the Origin and Evolution of the Universe	71
Chapter 6	Concluding Remarks	83
	List of References:	90
Appendix 1	Ptolemy's Theory	93
Appendix 2	Schrödinger's Cat	95
Appendix 3	On Space and Time	97
Appendix 4	Reality of Simultaneity	101
Appendix 5	Newton's Bucket	105
Appendix 6	Correction to Flawed Physics Lesson Regarding the Relativity of Simultaneity and Time	107
Appendix 7	Analysis of the Derivation of the Lorentz Transformation Providing Insight into the Special Theory of Relativity	115
Appendix 8	Einstein's Final (and Successful) Thought Experiment	123

AUTHOR'S PRELIMINARY COMMENTS

I add these comments in order to emphasize several points very important to me. I have expressed myself very strongly in parts of this treatise because of the frustration I felt upon reading some of the apparent nonsense written in books on relativity and quantum physics for the lay public. I have put a lot of time and thought into this difficult and controversial exercise and wish the end result to be positive.

Albert Einstein's Theories of Relativity have been called "the most beautiful scientific theory(ies) in the history of the world" (1d) and quantum theory has been touted, over and over again, as the most successful scientific theory ever. Scientists regularly claim that quantum theory led to the development of important modern technologies such as semiconductor electronics, lasers, MRI scanners, etc., etc. These do in fact rely on the understanding of the existence of quanta, the puzzling apparently dual particle/wave nature of matter, and the equations of Relativity and quantum mechanics. The standard interpretation of these theories and their equations, however, lead to a number of absurdities recognized by some physicists (see, for example, 2 and 40) but ignored by many. I discuss these absurdities in some detail in this book and provide alternative interpretations that may resolve much of what I have described as absurdities. I don't see the technical advances mentioned above as confirming the reality of the absurdities I have reviewed; rather, I believe that only the successful development of quantum computation, as it has been proposed, may provide some support for a reality behind one or two of these seeming absurdities.

If I am right in some of what I propose, verifying my theories will be positive and a contribution to science. And, where I am wrong, I will still consider it positive and a contribution to science as long as my proposals are proved to be in error and this is explained clearly with facts and logical argument, rather than ignored or explained with complicated mathematics that even the experts sometimes don't fully understand. There are thousands of professional physicists out there who are smarter, better trained and more experienced than I, so I expect one or both of these outcomes to be realized. To have some of these glossed-over issues properly discussed and explained will make this effort worthwhile.

PREFACE

When I was a very little child, I used to ponder the mystery of my existence and that of the universe. Why was all this here and how could I be part of it? I somehow knew that the occasion of my existence in the universe was an astoundingly low-probability event. My puzzlement at the existence of the universe itself was just as great but was tempered by realizing the even greater improbability of its *not* existing. It seemed obvious something had to exist. Space exists: we sense it; we exist in it. We can conceive of matter not existing – the disappearance of matter leaves empty space, but it seems impossible to conceive of space not existing. What would take its place? Empty space is nothing – if you took it away, nothing would be left, which of course is empty space.

I was overcome with wonder. Why was I here? How could I possibly be here? And what was this all about? Looking up at the star-filled sky (at that time the sky above my home in Manitoba, Canada, was an amazing profusion of countless bright twinkling objects), I couldn't help also pondering the concept of infinity. It seemed that the sky couldn't possibly go on forever; there must be an end. But if there was eventually an end, a boundary beyond which it couldn't extend, what was on the other side of the boundary? And the disturbing mathematical concepts of the curvature of space and finite unbounded manifolds (once I learned about them) did nothing to solve this conundrum, in my opinion.

I believed that gaining a strong intuitive grasp of the concepts of infinity and nothingness (infinite nothingness), as they relate to the origin, existence and eventual fate of the universe, would be important to begin grappling properly with this great puzzle. Of course, there has been, is and must be, an infinity of nothingness, into which something material needn't necessarily arise, but nevertheless has arisen and evolved into something with the ability to observe and ponder what it observes.

Sitting in the closet under the stairs in our little house on Gertrude Street, puzzling over these questions, I felt myself shrinking to an infinitesimal size then expanding to an enormous size, oscillating rhythmically back and forth between these unusual feelings. I never understood these weird oscillatory feelings, but as time went by, I eventually gravitated towards the study of science and got a grounding in physics. I now suspect there was something significant and fundamental about those weird sensations I felt as a child. I learned that science in general, and physics in particular, had made great progress against blind belief, superstition and mysticism over the last several hundred years. But it seemed to me there were still problems with the present state of modern physics theories, or at least the interpretation and lay presentation ("marketing") of them.

As I continue to ponder these problems today, particularly the outlandish wizardry inherent in modern-day quantum mechanics (at least as espoused by some enthusiastic and eloquent proponents), I can't help but appreciate the simple logic of those youthful musings. I hope I can expand on those thoughts to lessen the present confusing and disturbing muddle. I will try to elucidate this muddle, as I see it, in the first chapter, before proceeding with new thoughts on the matter.

I obtained a degree in physics fifty years ago and thought that after Einstein's contributions, and those of many brilliant folks before and after him, there was nothing of importance left to do in physics, particularly by anyone of lesser calibre than Einstein (Of course, I was completely wrong, not about relative calibre of course, but that nothing of importance was left to do). So, I moved into biophysics and eventually obtained a PhD in neuropharmacology. I spent many decades developing neuro-pharmaceuticals and medical devices and founding neuroscience start-ups, while retaining my interest in physics.

The advances in astronomy, cosmology and astrophysics during this period have been amazing and awe-inspiring. But during the last five to ten years I have been voraciously following the many interesting publications for the lay public by renowned physicists like Lawrence Krauss, Neil Turok, Brian Greene, Michio Kaku, and many others – and some of this stuff upset me. Don't get me wrong – these guys are all highly intelligent, trained physicists, and talented and imaginative writers, but some of the stuff they were propounding sounded to me more like String Theology, Quantum Mysticism and Parallel Confusion than bona-fide scientific theory and mechanics.

Of course, the genesis of this "stuff" goes way back to the early 1900s and the origins of the Theories of Relativity and quantum mechanics. Albert Einstein spent the last thirty to forty years of his life fighting the mysticism inherent in the Copenhagen Interpretation of quantum mechanics ("observation creates reality" – this is discussed further in the following chapters). Einstein seems to have lost that argument as the majority of physicists since have apparently accepted the Copenhagen Interpretation. It has even been suggested that Bohr felt that Einstein was "out to lunch" regarding his concerns about the general interpretation of quantum mechanics. However, I believe and demonstrate the exact opposite (see Appendix 8).

There is apparently a widespread belief that the Copenhagen Interpretation has been proven. We constantly hear the argument that quantum mechanics has been applied for many decades and is the most successful theory ever. There are good reasons for such a statement, and I accept these, but I am not ready to discard Einstein's concerns or ever accept that he was "out to lunch" – I have some of the same unresolved concerns he had until the end of his life. I am firmly on the side of Einstein, i.e., that he was right all along in his concerns, and will present a different way of looking at these things. I will attempt to show that although quantum mechanics has been successfully applied for many decades, that does not mean that the mystical Copenhagen Interpretation and all the new seemingly mystical concepts and theories that have evolved since then are on the right track.

Ironically enough, I find that Einstein himself may be partly responsible for some of the "spooky stuff" he objected to. I believe that these problems stem to some extent from the way he expressed certain things in his Theories of Relativity and how these theories have been interpreted. I will show some modified ways of expressing and interpreting these aspects of the relativity theories that eliminate some of the "spooky" mysticism. I will also present a different way of looking at the universe and the application of quantum mechanics that produces a different, possibly more logical, view of reality in the universe (as opposed to the observer-generated spookiness of the Copenhagen Interpretation, which may have some utility, but doesn't adequately enlighten us to the reality of nature or the nature of reality). Finally, I will propose a revised theory regarding the origin, evolution and nature of the universe that I believe may provide an explanation for "dark energy". And I should mention that, unlike the quantum absurdities that I discuss, this theory is falsifiable.

ACKNOWLEDGMENTS

I am eternally grateful to Brian Greene and Michio Kaku for so inciting me with their elaborate presentations of the seemingly mystical absurdities of quantum mechanics that I became determined to do something about it and attempt to clarify some of the absurdities. I am also grateful to Brian and Michio, as well as Lawrence Strauss, Neil Turok, Bruce Rosenblum, Fred Kuttner, Marcus Chown, Martin Bojowald and several others for their eloquent and readable accounts regarding relativity, quantum mechanics and related theories and concepts. I am also grateful to Manjit Kumar for his very interesting and lucid account in *Quantum: Einstein, Bohr and the Great Debate About the Nature of Reality*. Finally, I am indebted to a brilliant young Icelandic physicist, and a brilliant older physicist recently retired from CERN, who have generously reviewed and commented upon some of my treatise to challenge me and try to keep me from making too many egregious mistakes. Their role was important in preparing me for pushback from physicists irate at my ignorance and audacity. They are not responsible for any of the errors I may have retained in this treatise. Regarding anywhere I may be wrong in my analysis and assertions, I will not appreciate simply being attacked or dismissed for this, but rather will welcome reasoned arguments proving me wrong (and hopefully right, occasionally) and better explaining the issues.

CHAPTER 1

THE QUANTUM ENIGMA

1A INTRODUCTION

Science has come a long way in overcoming blind belief and superstition. The church was once the main source of information transfer and the information transferred, aside from the gossip among the masses that gathered together every Sunday, was hard-wired dogma not amenable to evaluation, criticism or change. It was still much this way in 1665-66, the "annus mirabilis" years of young Isaac Newton, who spent eighteen months during those years at his family home in Woolsthorpe because the plague had closed Cambridge University. During that time, he developed his laws of motion, the basis for what is now called "classical" mechanics. He had previously developed the powerful tool of calculus that has been so important for the evolution of "modern" physics.

Newton's "classical" mechanics was a stroke of genius that allowed freethinkers to begin to comprehend and control nature and the world around them, as well as start to grasp "real" truth beyond the dogma of "revealed" truth. Like all who make such major advances, Newton stood on the shoulders of "giants" (earlier geniuses) before him – these include Epicurus, Lucretius, Democritus, Bruno, Brahe, Kepler, Copernicus and Galileo, to name a few of the most notable. Bruno in particular, but also Galileo, was treated terribly by the Catholic Church and its horrendous Inquisition. Newton was a Christian, but apparently a rather unorthodox one.

Information access and transfer has of course exploded with the relentless march of 20th- and 21st-century information technology. But this has not quelled humanity's tendency to perpetuate dogma, nor its fascination with supernatural mystery. The successive generations of physicists over the last century seem to have accepted without much question the dogma of the Copenhagen Interpretation of quantum physics, in spite of all the mysteriousness it harbours. This is presumably because quantum mechanics appears to work unfailingly when applied to their problems. In the following, I will try to explain why this is so and why it may be misleading.

Newton's laws of motion and the "classical" mechanics they initiated burst on the scene as the first basic universal laws of nature. I believe that these are still the only basic laws of nature and that the Theories of Relativity and quantum mechanics have not really displaced them, but, rather, have merely rephrased them to correct for certain very special circumstances. Newton simply did not have the information available to him during his time to be able to recognize and deal with these special cases. Einstein acknowledged this when he received acclaim after the supposed experimental verification of his General Theory of Relativity.

I have great respect for Albert Einstein and he is one of my personal heroes, so my attempt to correct what I perceive as shortcomings or problems with the Theories of Relativity and quantum mechanics is not meant to criticize or demean him or his work in any way. Indeed, Einstein's physical intuition led him to seriously question the standard interpretation of quantum mechanics, and I will address this issue in detail in this dissertation. Similarly, although I, like Einstein, question the validity of the standard interpretation (i.e., the Copenhagen Interpretation), this is in no way meant to demean the important contributions to physics of Niels Bohr and his collaborators, nor of the many brilliant physicists since.

It is ironic that the way Einstein expressed and promoted some of his ideas regarding relativity may, in my opinion, have contributed to some of the very issues that bothered him. I will attempt to demonstrate this and to clarify some of these issues. I will further provide a revised interpretation of aspects of Relativity and quantum mechanics, attempt to remove the apparent absurdities, and possibly throw some new light on the nature and origin of the universe.

I am tackling these issues from the point of view of lay presentations published by many prominent physicists, including Einstein himself. Therefore, I use no complicated mathematics and most of my arguments should be understandable to non-physicists as well as physicists. While I am challenging some of the basic premises of modern physics, I do not mean this to reflect any lack of respect for the field and its practitioners. In fact, I have the greatest respect for science in general, and physics in particular, and I hope what I present will engender a positive discussion of these issues that might eventually advance the field.

While it is claimed that many, if not most, physicists usually just "shut up and calculate" (a statement often attributed to Richard Feynman but probably first used by N. David Berman), the absurdities are obvious and dramatic, and many physicists recognize them. Unfortunately, none of these absurdities have been adequately explained in understandable ways. I have tried to find logical explanations that do not require exotic mathematics that defy understanding or visualization. Everything I suggest in this treatise appears to logically resolve the issues tackled – I hope and expect that this will give experts the opportunity to either agree with my propositions, or disagree and explain how I am wrong. This process may provide a better understanding of the issues than currently exists. And to be useful, this must be done with clear logic that does not rely solely on exotic mathematics.

In the event I have misunderstood or misrepresented any of these issues, I look forward to being corrected. Many of these issues have been ignored or glossed over for decades and so at the very least, I hope that my efforts herein will expose these conundrums to the attention they deserve.

1B   THE ENIGMA

Quantum mechanics has been touted as the most successful theory ever devised and is used daily by physicists and engineers the world over. But, despite this degree of practical success, quantum theory is plagued by logical absurdities referred to as the "quantum enigma" (2) that I will address herein. As stated by the editors of Scientific American in their September 2015 issue (1a), which focused on Einstein and his accomplishments: "… a century of toil by generations of physicists has not accomplished a single theory of nature. Relativity and quantum mechanics are just as incompatible as they ever were."

The first most basic (and surprisingly broadly accepted) quantum absurdity is the interpretation that observation creates reality. That is, observation collapses the waveform representing an object and thereby creates reality — and that before observation, the "object" was a nebulous probability waveform existing everywhere but nowhere. This obviously can't be correct and I will try to clarify it below for those who need further explanation. I would be less certain and obstinate in this if there were any physical mechanism whatsoever proposed to explain this and the other commonly accepted absurdities produced by the present interpretation of the mathematics.

As mentioned earlier, many of the "absurdities" have emanated from the Copenhagen Interpretation (2, p. 100), captured in these statements: "… *until an observation or measurement is made a microphysical object like an electron does not exist anywhere.*" And "*It is only when an observation or measurement is made that the 'wave function collapses' as one of the 'possible' states of the electron becomes the 'actual' state and the probability of all the other possibilities becomes zero.*" (27, p. 219)

I will try to demonstrate the absurdity of these statements, and provide an alternative interpretation, throughout the course of this book.

Here are a few quotes from various sources that are based on the Copenhagen Interpretation developed and promoted by Niels Bohr, Werner Heisenberg and Wolfgang Pauli:

- *"... but the atoms or elementary particles themselves are not real; they form a world of potentialities or possibilities rather than one of things or facts."* (2a)
- *"... only the observed properties of microscopic objects exist."* (2b)
- *"No microscopic property is a property until it is an observed property."* (2c)
- *"The wavefunction is the only thing that physics describes – it's therefore the only physical thing."* (2d) Note: if this is correct, then quantum physics is definitely incomplete.

These statements at least focus on the microscopic or atomic, unlike some extensions of these mystic assertions, made mostly in physics publications for the lay public. Here is a statement in Neil Turok's *The Universe Within* (3, p. 93) that applies the concept of "observation creating reality" to the world: *"Before we observe it, the world is in an abstract, nebulous, undecided state. It follows beautiful mathematical laws but cannot be described in everyday language. According to quantum theory, the very act of observing the world forces it into terms we can relate to, describable with ordinary numbers."*

Again, I must make clear that I admire and respect Neil Turok, and enjoyed his book very much, but his statement illustrates the absurdity that I object to, on the macroscopic level this time. However, to give Turok his due, he does supply what could be construed as a better interpretation of the "standard" theory – i.e., **observation does not create reality but puts it into terms describable by ordinary numbers** – and this is close to an interpretation I will build upon. As I will try to demonstrate throughout this treatise, observation does not and cannot "force" the world, or reality in general, into anything – it merely puts numbers we think we can understand into our very limited picture/understanding of the real world. In other words, **observation creates an "observed world", which is an imperfect subset of the real world.**

Regarding the status quo for quantum mechanics, let's remember that just because something seems to work does not necessarily mean that it embodies the full truth. Consider this example quoted from *Quantum Enigma* (2e): *"In the 2nd century AD Ptolemy of Alexandria described the heavenly motions so well that calendars and navigation based on his model worked beautifully. The astrologer's predictions – at least regarding the positions of the planets – were likewise accurate."* But, of course, Ptolemy's theory (Appendix 1) was totally false.

Indeed, while many physicists today claim that quantum theory is the most successful theory yet devised, based on their application of its equations to various problems, I believe that I can show that many of its implications regarding quantum "reality" are not just weird and unintuitive, but very misleading. These don't represent the "real" world, but rather a **"quantum observer's" world.**

The quantum theory concept of a particle having no specific position, but being everywhere or nowhere until observed, is obviously just not so. Astronomers and astrophysicists are constructing pictures of the universe that have been transmitted unobserved for billions of years, but the photons have not strayed from their paths in all that time and distance. It is clear that particles (in this instance photons) have specific positions and velocities and can travel in straight lines "forever" (and indeed must do so until they encounter a force or field that changes their speed or direction, as Newton's first law correctly taught us), all without being observed. Otherwise, if quantum mechanics was correct (in the sense of particles having no specific defined positions or velocities, and not existing until observed), the pictures we receive would show nothing, or at best would be garbled nonsense – but they are not. A clever debater for nonsense might try to argue, "But the astronomers are observers and they are observing millions, even billions, of years into the past and so are creating the past reality." I won't waste the required paper space to shoot down such mystic nonsense.

And to be clear, Heisenberg's Uncertainty Principle states that we can't *determine*, beyond a specified limit of accuracy, both the position and velocity of a particle (observation), not that a particle can't *have* a definite position and a definite velocity (reality). If a particle, or object of any sort, were not real, or with "position", there would be nothing to observe. Observation doesn't create reality; rather, it gathers a piece of information to put into the "observer's" limited reproduction of reality. The wave equation isn't some magical reality descended from a "creator" in "heaven"; it is simply an equation developed by a brilliant but mortal classical physicist that presents the full probabilistic spectrum of where one might find a particle. Upon observation, the highest-probability position is confirmed and entered into the observer's notebook to become part of the "observer's world", i.e., the observer's attempt at reproduction on paper, or in a computer database, of a part of the "real world". At this point, the rest of the positions or "states" presented by the equation are confirmed to be meaningless. **The only thing that "collapses" is the impression in the mind of the observer that there was reality to any of those other "states".**

Furthermore, let's consider the evolution of the universe and ultimately our world and its inhabitants. None of this could have happened if particles, and their products, had no position or properties and did not exist until observed – in fact, observers did not exist until the above-mentioned evolution had progressed for billions of years. Standing in front of a standard quantum theory believer, one would have to say that: **"If you are right then you cannot possibly exist – therefore you cannot possibly be right!"** And let's not get sidetracked with incoherent (or should I say decoherent) nonsense such as that atmospheric molecules might act as "observers" (4, p. 210). According to quantum theory, these molecules also would not exist until observed, and so this nonsensical theory tracks back to nowhere.

Attempting to understand how brilliant physicists could come to such an impossible conclusion as the Copenhagen Interpretation, I thought of Bertrand Russell's discussion of inductive versus deductive reasoning in *A History of Western Philosophy* (first published in

1945, Simon and Schuster, NY). Inductive reasoning based on observation is the hallmark of the scientific method that has advanced society so far beyond the earlier superstitious eras, which were driven by the "revealed word". Thus, it seems that the quantum theorists have been reasoning inductively based on what they (or their experimental counterparts) have observed, but because it is impossible to measure everything, we must use deductive reasoning to fill in the gaps to reveal and understand reality. For example, let's suppose a particle is measured at position A and then at position B. Then, since it wasn't measured at any position between A and B, by the standard interpretation it is claimed that the particle doesn't, or didn't, exist anywhere between A and B. But of course, with the use of deductive reasoning, we understand it must have existed at every position between A and B. (Note: to the purists in the crowd this example may be a stretch, but it makes the point without getting too complicated.)

Rosenblum and Kuttner, in their book *Quantum Enigma* (2, p. 85), state that: *"Engineers and physicists who work with (these technologies) may deal intimately with quantum mechanics on an everyday basis, but they need not face up to the deeper issues raised by quantum mechanics. Many are not even aware of them. In teaching quantum mechanics, physicists, including us, minimize the enigmatic aspect. We don't distract students from the practical stuff they need to use. We also avoid the enigma because it's a bit embarrassing; it's been called our skeleton in the closet."* They also state through a fictional physicist in their treatise: *"But no prediction of the theory has ever been wrong. … Whether or not it fits our intuitions should be irrelevant."* I strongly disagree with this very common contemporary attitude – if something doesn't agree with our intuition, or experience, gathered from decades of observations, we should question why and search vigorously for an answer. This prevailing attitude explains a lot, and herein lays a potential solution to breaking out of this century-old logjam. **The issue is not whether the equations work; it is "why do they work" and "what does this really mean"?**

As I mentioned earlier, many of the "converted" seem to think that the wave equation was delivered to earth by some "creator" with the guarantee that all the waves it mathematically describes actually occur in nature and represent the only reality. But, in true reality, they are simply mathematical equations developed to try to predict reality by a smart guy named Erwin Schrödinger, who himself, like Einstein, could never accept the mystic hocus-pocus accepted by so many "quantum physicists". Indeed, he even put forward his "cat conundrum" to show how ridiculous the quantum superposition mysticism was; but even this tongue-in-cheek exercise has been taken over and twisted by some to promote the possibility of superposition at the macro level. I address this maddening nonsense in Appendix 2.

Rosenblum and Kuttner in *Quantum Enigma* (2) provide some insight into the forces working for decades on the minds of gullible/vulnerable physicists. Here are a few of their enlightening statements: *"… then came the politically and socially 'straight' 1950s. In physics departments a conforming mind-set increasingly meant that an untenured faculty member might endanger a career by seriously questioning the orthodox interpretation of quantum mechanics. Even today it's best to explore the meaning of quantum mechanics only while also working a 'day job' on a mainstream physics topic."* (p.139); *"In 1965 when Bell's Theorem was published, it was a mild heresy for a physicist to question quantum theory or even to doubt that the Copenhagen interpretation settled all the philosophical issues."* (p. 148); *"... in the early 1970s investigation of the fundamentals of quantum mechanics was not yet considered proper physics in most places."* It was said about a physicist applying for a professorship, *"What has he done besides checking quantum theory? We all know it's right."* (p. 150). This sort of stuff makes physics sound more like political dogma or a religion than a science. This disturbs me and has incentivized me to try to sort through the issues and come up with a more realistic interpretation.

I will try to make the case that the Theories of Relativity and quantum mechanics have not been so successful because they represent the "real" world but because they reproduce, or should I say accurately predict, what will be observed; in other words they faithfully represent the "observed" world. There is only one "real world", but an infinite number of potential "observed" worlds, each requiring a correction for "point of view", or should we say "rigid coordinate system", which is what the Theories of Relativity and quantum mechanics provide. This may be why the application of the Theories of Relativity and quantum mechanics to certain important theoretical questions often yields an infinite number of possible solutions, giving rise to suggestions of multiple dimensions and multiple universes, and to singularities as well. The explanation for these absurdities is probably that these theories incorporate an infinity of possible "rigid" coordinate systems, each with a different point of view.

Thus, observation creates the "quantum world", but the concept that "observation creates reality" is impossible to accept. Most will agree with this assertion regarding the macroscopic world in which we live and observe. But simple logic assures one that this absurd concept cannot apply to the microscopic world (photons, electrons, atoms, other particles, etc.) either, or "observers" would never have evolved. For billions of years these various particles, from quarks to atoms to molecules to dust to stars to planets and beyond, have existed without observation, and evolved eventually into conscious creatures who could observe and question what they observed. I will attempt to show later how consciousness may in fact come into the equation and might supply the only possible additional dimension(s) beyond the obvious four of very real and absolute space and time.

Brian Greene has been one of the major quantum physicists promoting quantum mysticism to the lay public, so I will review and dissect some of the assertions he has made in his lay publications. Before beginning, let me acknowledge that Brian is a highly respected physicist and a far more eloquent writer than I am. I admire him for this and thank him for getting me upset enough with his quantum assertions to try to come up with alternatives to the silliness I perceived.

Here is a list of statements in *The Fabric of the Cosmos* (4) that I will try to deal with logically:

p. 6: "... *space and time top the list of age-old scientific constructs that are being fantastically revised by cutting-edge research.*"

In my opinion some of these "fantastic" revisions need re-revision, as I will attempt to elucidate herein.

p. 9: "*While struggling with puzzles involving electricity, magnetism and light's motion, Einstein realized that Newton's conception of space and time, the cornerstone of classical physics, was flawed. ... he determined that space and time are not independent and absolute, as Newton had thought ... Some ten years later, Einstein hammered a final nail in the Newtonian coffin by rewriting the laws of gravitational physics.*"

Great work was done but no final nails were driven and no new "laws" were written. Furthermore, in my opinion it is the new "conception of space and time" that is flawed, not Newton's original conception.

pp. 9-11: "*This time, not only did Einstein show that space and time are part of a unified whole, he also showed that by warping and curving they participate in cosmic evolution.*"

I argue in Chapter 2 that this is not a correct interpretation of the mathematics of the General Theory. Warping or curving of space doesn't explain gravity, and indeed can't possibly occur – instead, I will offer a different interpretation of the General Theory of Relativity.

"*The two theories of Relativity are among mankind's most precious achievements, and with them Einstein toppled Newton's conception of reality. ... Ours is a relativistic reality.*"

I argue later on that these important "precious achievements" toppled nothing – but merely added a useful correction to Newton's reality. Reality is not relativistic, only our **observation** of reality is.

"*Yet, because the deviation between classical and relativistic reality is manifest only under extreme conditions (such as extremes of speed and gravity), Newtonian physics still provides an approximation that proves extremely accurate and useful in many circumstances. But utility and reality are very different standards. As we will see, features of space and time that for many of us are second nature have turned out to be figments of a false Newtonian perspective.*"

This begins as a refreshing acknowledgment, but I will show further on that this last sentence is not quite true – it is the new "perspective" that I believe is flawed.

"*The second anomaly to which Lord Kelvin referred (aspects of the radiation objects emit when heated) led to the quantum revolution, one of the greatest upheavals to which modern humanity has ever been subjected. By the time the fires had subsided and the smoke cleared, the veneer of classical physics had been singed off the newly emerging framework of quantum reality.*"

Planck's quantum discovery (subsequent to Lord Kelvin's observations) was real and important but the last sentence here is dramatically exaggerated. I can't believe the hyperbolic fervour with which the quantum enthusiasts enjoy dissing classical physics. **The "quantum world" is not the "real world", but the "observed world", i.e., the "real world" corrected for observational error (so that what is observed matches what is predicted).**

The editors of Scientific American (1a) have used similar overly dramatic descriptions for the significance of the Theories of Relativity and quantum mechanics. These exaggerated words reflect the sensational headlines in the *Times of London* in 1919, when Eddington reported a rather weak confirmation of Einstein's so-called relativistic ideas: "Revolution in Science – New Theory of the Universe – Newtonian Idea Overthrown". Einstein himself showed much more humility in his statements at the time. In contrast to the boastful headlines at the time, he stated (41, p. 48): *"Newton forgive me. You found the only way which, in your age, was just about possible for a man of highest thought and creative power."* Of course, there is a slight suggestion in these words that he thought that he had transcended Newton's laws.

In fact, Einstein didn't rewrite the laws of gravitational physics, but derived more accurate equations, providing a correction for a special circumstance and suggesting a new interpretation of gravity as operating by warping or curving space. I will show in Chapter 2 that this interpretation of gravity is likely incorrect and will provide what I believe is a more reasonable interpretation of what Einstein accomplished here. I will further argue that time and space are in fact real and independent, and that there is a good non-mystical reason they cannot be assigned absolute values except relative to a surrogate "rigid" coordinate system.

p. 11: "Without equivocation, classical physics declares that the past and the future are etched into the present. … But according to the quantum laws, even if you make the most perfect measurements possible of how things are today, the best you can ever hope to do is predict the probability that things will be one way or another at some chosen time in the future, or that things were one way or another at some chosen time in the past. The universe, according to quantum mechanics, is not etched into the present; the universe, according to quantum mechanics, participates in a game of chance." I consider the implications of this last statement to be wrong-headed and misleading, as did Einstein.

There is certainly a strong component of chance or probability in the progression of events in the universe, but not because determinism and cause and effect are not a part of reality, or because the past and the future are not represented in the present. In fact, the apparent indeterminism is more likely the result of the often-overwhelming number of unknowns, as well as the acknowledged imperfections in, and consequences of, our observations or measurements. Chaos Theory demonstrates how infinitesimal errors in observation (measurement) can lead to catastrophic changes in predicted outcome (5).

"Although there is still controversy over precisely how these developments should be interpreted, most physicists agree that probability is deeply woven into the fabric of quantum reality." I will attempt to deal with this controversy by providing a more reasonable interpretation and show that there is no such thing as a "fabric of quantum reality" (more likely a "quantum fabrication of reality").

"… quantum mechanics describes a reality in which things sometimes hover in a haze of being partly one way and partly another. Things become definite only when a suitable

observation forces them to relinquish quantum possibilities and settle on a specific outcome." Yikes, what a disturbing use of the physical concept of "force"! If physicists could demonstrate the existence and nature of this "force" I would be suitably impressed and chastened.

In actuality, the wave equation presents the imagined possible world in the equations and consciousness of the observer, and the observation merely confirms the highest probability and thereby collapses all the other possibilities residing in the observer's mind/equations. Things become definite in the observer's mind and notebook when a suitable observation is made. I argued some pages back that the concept that "things" or "reality" hover in a haze of being partly one way and partly another is impossible nonsense.

pp. 11-13: "... *time ... seems to have a direction pointing from past to future – for which neither relativity nor quantum mechanics has provided an explanation. ... But where does time's asymmetry come from? ... It turns out that the known and accepted laws of physics show no such asymmetry: each direction in time, forward and backward, is treated by the laws without distinction. ....., special physical conditions at the universe's inception (a highly ordered environment at or just after the big bang) may have imprinted a direction on time ..."*

What this actually shows is that there is still a serious problem with these laws – they are not yet complete. I will further on provide a possible explanation for some of the phenomena mentioned. <u>Nothing imprinted a direction on time. Rather, time imprinted a direction on reality and the evolution of the universe.</u>

p. 14: "*Its [the Big Bang's, 1960's] successes notwithstanding, the theory suffered significant shortcomings. It had trouble explaining why space has the overall shape revealed by detailed astronomical observations, and it offered no explanation for why the temperature of the microwave radiation ... appears thoroughly uniform across the sky. Moreover ... the Big bang theory provides no compelling reason why the universe might have been highly ordered near the very beginning, as required by the explanation for time's arrow.*"

I appreciate these admissions regarding the problems with the Big Bang Theory. My approach will explain the uniform temperature of the microwave radiation, and the shape of the universe near the beginning. It will also explain the so-called direction of time and the independence of time and space, and clarify the issue of creation as it applies to time and space. There is no need for "high order at the very beginning" to explain time's arrow – time has no "arrow" and there was no beginning of time or space. Time may not be properly understood yet, and I will try to explain this later on.

pp. 14-15: "*Inflationary cosmology [late 1970's/early 1980's] modifies the big bang theory by inserting an extremely brief burst of astoundingly rapid expansion during the universe's earliest moments (in this approach, the size of the universe increased by a factor larger than a million trillion trillion in less than a millionth of a trillionth of a trillionth of a second). ... Yet ... for two decades inflationary cosmology has been harboring its own embarrassing secret. Like the standard big bang theory it modified, inflationary cosmology rests on the equations Einstein discovered with his general theory of relativity. Although volumes of research articles attest to the power of Einstein's equations to accurately describe large and massive objects, physicists have long known that an accurate theoretical analysis of small objects – such as the observable universe when it was a mere fraction of a second old – requires the use of quantum mechanics. The problem, though, is that when the equations of general relativity commingle with those of quantum mechanics, the result is disastrous. The equations break down entirely ..."*

I appreciate these admissions as well, since Inflationary Cosmology is even more unbelievable than the Big Bang Theory. The above will be clarified by suggesting that the Big Bang and Inflation theories are not correct, i.e., they never occurred as described. The Relativity equations are being misinterpreted and misused, and, furthermore, the standard interpretation of quantum mechanics doesn't seem to make any sense.

And then we have EPR, quantum entanglement, decoherence, Bell's theorem and Bell's inequality, String Theory, M-theory, parallel worlds, multiple universes, etc., etc. Brian Greene's (e.g., 4) and Michio Kaku's (e.g., 6) excessively reverent treatment of these magical phenomena were mainly what upset me enough to tackle these concepts, along with the whole mystical shroud covering the impressive successes of quantum mechanics. Brian Greene (4, p. 114) states *"Pairs of **appropriately prepared particles** [implying as he states openly and constantly elsewhere that "the details don't matter"] – they're called entangled particles – don't acquire their measured properties independently. They are like a pair of magical dice, one thrown in Atlantic City and the other in Las Vegas, each of which randomly comes up one number or another, yet the two of which somehow manage always to agree. Entangled particles act similarly, except they require no magic. Entangled particles, even though spatially separate, do not operate autonomously."* Entangled particles may not operate autonomously or acquire their measured properties independently, but the magical dice described are not a good example of such apparent non-magic, non-autonomous, non-independent entanglement.

In fact, before swallowing this "non-magic", we need to carefully analyze the "unimportant details", including the "appropriate preparation". Even more importantly, we need to hear a mechanism for these "non-magical" magical things. I don't agree with the "standard" take on these various quantum phenomena. Greene does suggest that the acceptance of what he describes is not unanimous and there may be other ways of looking at these things, but he claims he is not on the side of the doubters (I certainly am). My purpose is to explore some other ways of looking at these things. I will tackle quantum entanglement, decoherence and Bell's Inequality (as an argument against reality and locality) in Chapter 3 and proceed now to deal with other issues in the so-called relativistic and quantum worlds.

Meinard Kuhlmann in his August 2013 article in Scientific American (7) clearly identified some of these problems with quantum mechanics but then proceeded, I believe, to muddy the waters even further. In his article, Kuhlmann states, *"... physicists are not even sure what the (prevailing) theory says – what its 'ontology', or basic physical structure, is. ... The unsettled interpretation of quantum field theory is hobbling progress .... It is perilous to formulate a new theory when we do not understand the theory we already have."*

I am in some agreement with these statements (although I think that if you can't understand the theory that you have, you should try to develop a new theory that you can understand), but Kuhlmann in his brief bio also states, referring to his behaviour as a student, *"I would ask a lot of questions just for fun and because they produced an entertaining confusion."* Kuhlmann then proceeds to provide the same "entertaining confusion" in the remainder of his article. In the remainder of this section, I will attempt to clarify Kuhlmann's entertaining confusion and then address some of the entertaining confusion that physicists have been building around classical versus quantum physics.

Kuhlmann proceeds to make statements like *"… a particle inside your body is not strictly inside your body. An observer attempting to measure its position has a small but nonzero probability of detecting it in the most remote places of the universe. … Relativistic particles are extremely slippery; they do not reside in any specific region of the universe at all."* And, *"… suppose you had a particle localized in your kitchen. Your friend, looking at your house from a passing car, might see the particle spread out over the entire universe* [**of course, this is not very likely to say the least**]. *What is localized for you is delocalized for your friend. Not only does the location of the particle depend on your point of view, so does the fact that the particle has a location."* And I guarantee that this is absolute rubbish based on a misunderstanding of what the wave equation is telling us.

In these pretty much status quo, but disturbingly outlandish, statements, Kuhlmann is simply playing on the whimsical "observer-dependence" interpretation in quantum mechanics that has fuelled so many popular lay publications by some physicists. However, this interpretation has also contributed to much of the "entertaining confusion" in contemporary physics.

Kuhlmann goes on to conclude that the problem is that *"the classical notion of a particle is misleading us; what the theory is describing must be something else."* He then considers fields but eventually concludes *"Clearly, then, the standard picture of elementary particles and mediating force fields is not a satisfactory ontology of the physical world. It is not at all clear what a particle or a field even is."*

Again, I don't disagree completely with these statements, but I believe that the standard picture of particles and fields is intuitively clearer for 99% of circumstances than the modern quantum interpretation is. Nevertheless, Kuhlmann goes on to suggest that *"a growing number of people think that what really matters are not things but the relations in which these things stand."* He labels this as *"epistemic structural realism"* and goes on to conclude *"the reason that we can know only the relations among things and not the things themselves …. is that relations are all there is"* and labels this concept *"ontic structural realism."* I consider all this to be merely more "entertaining confusion" and will try to demonstrate this by looking at some of Kuhlmann's simple everyday surrogate examples for this concept.

One example is the "mirror-symmetric face": Kuhlmann claims *"a mirror swaps the left eye for the right eye, the left nostril for the right, and so on. Yet all the relative positions of facial features remain. Those relations are what truly define a face, …"*.

This is obviously not quite right – very few, if any, faces are "mirror-symmetric", and even if they were the "relations" have no meaning without the presence/reality of the things they are relating. For example, what if we replaced the nose with a fire hydrant or maybe a tree? Even without such dramatic exchanges, the so-called mirror symmetric face does not reproduce the original – they are not identical.

Another example he uses is a subway network, and he claims: *"It is the structure of the network that matters primarily"*, implying that the reality of the stations is irrelevant. But, of course, the connections are again meaningless without the existence and identification of the stations. After all, do we not know where we are going when we travel? Since the "things" that are being "related" in both examples are potentially infinite in number, the reality and identification of the specific "things" is essential to give meaning to the "relations".

Kuhlmann then proceeds to suggest another approach giving priority to "properties": *"Many philosophers, including me,"* he says, *".... think it would be better to view properties as the one and only fundamental category. .... You can regard properties as having an existence, independently of objects that possess them."*

However, let's note that properties, or "tropes" as the philosophers have rakishly chosen to designate them, did not exist until consciousness evolved. Consider Kuhlmann's example of "redness" – in fact "redness" has no real meaning beyond the perception of "redness" by consciousness. The "real world" has existed since "the beginning", but the "observed world" has only entered the scene subsequent to the evolution of consciousness.

I applaud Kuhlmann's elucidation of some of the mysteries of the vaunted quantum physics, as well as his attempt to deal with the present dilemma in physics through new ideas and concepts. But "ontic structural realism" just doesn't seem to improve the situation, in my opinion. It was based loosely on the "observer dependence" interpretation rampant in quantum mechanics. This unfortunate interpretation is responsible for much of the confusion in contemporary physics, as I will continue to attempt to demonstrate in this treatise.

Indeed, I believe that one essential problem we need to deal with in confronting the quantum madness boils down to clarifying the distinction between the "observed world" and the "real world" that I believe has been twisted by the eloquent proponents of "quantum mysticism". This I will attempt to do after I have reviewed the contribution of the Theories of Relativity to the quandary and have discussed the problems with the Big Bang Theory, Inflation, String Theory, misconceptions in the "revolutionary new concepts" of space and time and the questionable experimental validation of these flawed concepts.

CHAPTER 2

CONTRIBUTION OF THE THEORIES OF RELATIVITY TO THE QUANTUM ENIGMA

As mentioned earlier, Albert Einstein did not agree with the Copenhagen Interpretation of quantum mechanics, which he felt was not a complete theory. He spent the last decades of his life battling with the Copenhagen Interpretation's proponents and searching for a more complete unified theory. It is sadly ironic that some of the absurdity in quantum mechanics seems to have had its birth in the misleading way in which Einstein's Theories of Relativity have been interpreted and presented, and even in some of the wording Einstein himself used in stating some of his conclusions.

I will first address a couple of widely held and exaggerated, if not fictional, beliefs regularly expressed in the media about relativity and quantum mechanics. I will then deal with some of my concerns with the generally accepted interpretation of the Theories of Relativity and with some of Einstein's statements that I believe were unintentionally misleading.

- It has been repeatedly stated, including in the Scientific American article (1b) touting Einstein's accomplishments, that quantum mechanics and the theory of relativity *"toppled Isaac Newton's physics and redefined our notion of space and time."*

This is the first piece of common/universal exaggeration. In fact, Newton's classical mechanics is still the only undeniably correct theory of nature that we have; it merely needs to be corrected for certain extreme situations because the speed of light is not infinite (i.e., information transfer is not instantaneous), which obviously must affect observations. And this is what Einstein's Special Theory of Relativity does, as I will demonstrate as we proceed.

- The attempts to **explain away Einstein's discomfort** with quantum mechanics should be replaced with attempts to **explain Einstein's discomfort**, which was very understandable in my opinion. I will argue as I proceed that he was right.

Einstein's widely recognized special strength was his physical intuition. Apparently, Einstein obtained help to develop the equations for his General Theory of Relativity and it was these equations that led to the touted "break-through" explanation of gravity as the result of the "curvature of space-time". This explanation is widely accepted by quantum physicists (not surprisingly since the equations seem to work), but, as I will show further on, this explanation is probably not quite correct. We should honour Einstein's intuitional prowess and put more energy into following up on his hesitations regarding quantum mechanics, as well as to finding more reasonable interpretations of the quantum absurdities – as I will attempt to do in the succeeding chapters.

2A   THE LIMITATIONS OF MATHEMATICS

Mathematics is not foolproof – it is a tool we use to facilitate our attempt to discover reality, but it does not always represent reality. Mathematics is a language, a powerful and very special language, but it can tell us lies as well as truths. While there have been many brilliant and talented physicists who have promoted the importance, even supremacy, of mathematics as a guiding principle to the elucidation of physical law, one must remember that there have also been many talented writers and speakers who have used language, with great eloquence, to promote great fictions as well as great truths.

Mathematics provides a step-by-step mechanism to progress logic, but the integrity of the outcome of this logical mechanism is dependent upon the validity of the input – i.e., the assumptions or premises and the data that are fed into the mechanism. This was most succinctly put by Kenneth Boulding in a lecture given at the University of Michigan in December 1955, in which he said: "*Any given mathematical process is a sausage-machine – what comes out is determined entirely by what goes in.*" And in my opinion, given some of the absurdities that have come out of the quantum mechanics sausage-machine, there must have been some questionable things fed into it. Indeed, the physics resides not only in the machine, but also in what is fed into the machine.

Paul Dirac once said, "*It is more important to have beauty in one's equations than to have them fit experiment.*" (15)

He provided an interesting explanation for this comment, which I found appealing, but this still seemed to be a disturbing statement for a scientist to make, particularly someone as brilliant and talented as Dirac (see also 37). To put this into context, let me again quote Boulding: "*The more mathematicians talk to each other the less skilled they become, very often, in communicating with the outside world. Knowledge monopolies ('mysteries') also tend to lead to high status in any culture, and there is a constant temptation of any priesthood towards a cult of unintelligibility. Mathematicians are peculiarly exposed to this temptation, they come to value style and elegance according to internal aesthetic patterns which have little to do with communicability* [or on occasion reality]."

There is probably some truth to this statement, but I always associated the above type of behaviour more with politics, religions or cults than with mathematics.

One of the major conceptual problems in physics that has probably led to mathematical deception involves the complicated issues presented as 1D or 2D (where D = dimension) simplifications. The problem is that 1D and 2D objects do not exist in reality – they are totally imaginary and can't represent real 3D objects or situations. An example in physics is the so-called "2D" balloon surface model for the expansion of the universe – I will discuss this further in another section.

I need to quote here Dr. Mahesh Jain (May 2012), who said: "*Scientists, particularly physicists, and others regard Mathematics as the final proof of anything. It is commonly held that once a proposition is able to find some mathematical argument in its support, it is deemed proved and settled for good. ... The current cosmic view in Physics has its origin in two fundamental hypotheses of Einstein's Theory of relativity, mathematically logically extrapolated to a theory of entire cosmos, directed towards reconciliation with quantum physics. ... Strings of the string theory and M-Theory are beyond sensory perception – direct or indirect. These efforts have led to theoretical models which are quite similar to Ptolemy's Model of Planetary System and the Phlogiston Theory of Combustion. This leads us to the question of natural limitations of mathematics in understanding natural events, things and the cosmos as a whole: 1) Mathematics is not all real: ... Mathematics is not like a magician's hat that can apparently produce things out of nothing. 2) Until and unless, one clearly knows what data is to be collected and how the relevant data is to be interpreted, how mere application of a mathematical tool or technique can ever lead to prediction of a natural event is clearly beyond comprehension.*"

I think Jain's last comment above may be a bit overstated, and I must intervene here to applaud all the impressive advances physicists and mathematicians have made over the last century, while acknowledging that some of the more mystically absurd concepts that invade quantum mechanics probably emanate from the inappropriate application and/or interpretation of mathematical results or implications. Mathematics is a powerful tool, but it is only that, a tool, and can predict nonsense. I will argue later on that the "curvature of space", touted as the explanation for gravity, is most likely the result of a misinterpretation of some complicated mathematics. But first let me provide some examples that demonstrate what I consider misleading wording that Einstein used to explain his reasoning in his

Special Theory of Relativity, and the flaw in the suggestion offered by others that this theory "toppled Isaac Newton's physics".

2B MISLEADING ASPECTS OF THE SPECIAL THEORY OF RELATIVITY

In his treatise on the Theories of Relativity (9, p. 61), Einstein states that *"The non-mathematician is seized by a mysterious shuddering when he hears of 'four-dimensional' things, by a feeling not unlike that awakened by thoughts of the occult. That we have not been accustomed to regard the world in this sense as a 4D continuum is due to the fact that in physics, before the advent of the Theory of Relativity, time played a different and more independent role as compared with the space coordinates."*

This strikes me as an immense exaggeration – this recognition does not require the Theory of Relativity. Indeed, this has always seemed intuitively obvious to me: it is easily pictured in the mind how time acts as a fourth dimension, not as a fourth space dimension, but as a completely independent fourth dimension that expands what the three space dimensions tell us. Motion in space occurs with time, but time is not a space coordinate. And I will argue later on that, in fact, "the fabric of spacetime" is not a real entity, but rather a **fabric**ated concept.

Einstein continues, *"As a matter of fact, according to classical mechanics, time is absolute."* And, as I will argue, yes, it is, and that doesn't interfere at all with the vision of it as a fourth dimension. Einstein's implication is that time is relative, not absolute, but <u>it is only the **observation/measurement** of time that is relative; time itself is not.</u> I believe that recognition of this will help advance modern physics.

Also, in his treatise dealing with "Special Relativity", Einstein (9, pp. 10-11) dismisses the concept of space, wrongly in my opinion. Regarding a situation he imagines wherein he drops a stone from the window of a moving train that he observes to *"descend in a straight line"* while a pedestrian on a nearby footpath *"notices that the stone falls to earth in a parabolic curve"*, he states: *"I now ask: Do the 'positions' traversed by the stone lie 'in reality' on a straight line or on a parabola? Moreover, what is meant here by motion 'in space'? From the considerations of the previous section the answer is self-evident. In the first place, we entirely shun the vague word 'space' of which we must honestly acknowledge we cannot form the slightest conception, and we replace it by motion relative to a practically rigid body of reference .... With the aid of this example it is clearly seen that there is no such thing as an independently existing trajectory (lit. 'path-curve'), but only a trajectory relative to a body of reference."* Yes, in fact there definitely is a trajectory relative to this body of reference. Indeed, there are an infinite number of such trajectories relative to an infinite number of such potential surrogate frames of reference – but the only **real** trajectory is the trajectory relative to **space** – this is the only true reality and unfortunately is the one that relativity and quantum mechanics have discarded.

Unfortunately, I believe Einstein's analysis above is somewhat misleading and provides a basic misconception that is causing some of the absurdities apparent in quantum mechanics, and the reason the Theories of Relativity and quantum mechanics become incompatible and can't extrapolate back to any reasonable "beginning". In fact, space is real and important and the "motion in space" is the essence of "reality". However, for reasons I will discuss elsewhere (Appendix 3), there is no origin to provide a benchmark in space for the observer, so the **observer** must adopt a "rigid" frame of reference as Einstein suggests – but a problem is then introduced as there are an infinite number of arbitrarily possible frames of reference, each of which must be corrected for depending on the relative position, motion and direction of motion between the observer and the observed. (This is what the Special Theory of Relativity does – it is not a new law of nature but a correction necessary to validate the observation from the chosen frame of reference.)

Throughout relativity and quantum mechanics, the "**real world**" is replaced by the "**observed world.**" Of course there is such a thing as an independently existing trajectory, and replacing it with a trajectory relative to a rigid body of reference allows a calculation of what will be observed with respect to that body of reference. However, this concept also replaces the singular reality with an infinite number of potential observational surrogates. And this explains the real meaning and import of the Copenhagen Interpretation. Quantum mechanics predicts an observed world, and this doesn't exist until observations are made – observation creates the observed world (a.k.a. the quantum world) and nothing (e.g., a particle) exists in this observed quantum world until an observation is made. But, of course, it does exist in the real world regardless of observation.

Einstein continues further in his famous treatise to claim: *"I allow myself to be deceived as a physicist (and the same applies if I am not a physicist), when I imagine that I am able to attach a meaning to the statement of simultaneity."*

He asks the reader not to proceed until he/she has accepted this statement, then proceeds to develop his argument against simultaneity. He presents an example using two "simultaneous" flashes of light as observed from a train moving with the velocity v relative to the embankment, compared to what is viewed with the embankment as the frame of reference. His observation that simultaneity is not observed from both frames of reference is correct and obvious, but his subsequent statement that *"every reference body has its own particular time"* is misleading (there is time and there is the perception of time). I repeat: It is only the perception of time that is relative – time itself is not.

The concept in Einstein's quote above would be more correctly stated as follows: *"every reference body provides its own assessment of the timing of events."* See Appendix 6 where I demonstrate the flaw in an exercise used to teach physics students about the relativity of simultaneity and time.

In fact, time is (and must be) progressing identically everywhere, but corrections must be made to the observation of "distant" events, depending on the actual distance and the relative velocities between the frames of reference. This is fully consistent with classical mechanics as well as relativity. We don't have to *"discard the assumption"* that *"time has an absolute significance"* to eliminate *"the conflict between the law of the propagation of light in vacuo and the principle of relativity"* or to assess the reality of *"simultaneity"*. (There is a flaw in Einstein's analysis of the conflict between the law of the propagation of light in vacuo and the Principle of Relativity, and I deal with this in Appendix 3.) <u>Einstein defines a real issue, but misrepresents its true meaning and significance. We are presented with a new "law of nature" that is nothing more than a correction for observation because the speed of light, which is required for observation, is finite rather than infinite.</u> I confirm this further on in this chapter.

In the following I will use several versions of Einstein's "two flash" example to show that simultaneity occurs <u>in reality</u>, and this can be easily demonstrated.

First let's look at three examples using a stationary observer:

1. Stationary Observer SO intermediate between simultaneous flashes A and B:

$$t_0 \quad\quad t \quad\quad t_0$$
$$B \text{------------ SO ------------ } A$$
$$X_1 \quad\quad\quad X_2$$

Without any effort it is easily seen that SO sees the simultaneous flashes as simultaneous. <u>$X_1 = X_2 = c(t - t_0)$, so simultaneity occurred and could be confirmed.</u>

2. Stationary Observer SO situated unequally between simultaneous flashes A and B:

$$t_0 \quad\quad\quad\quad\quad\quad\quad\quad\quad\quad t_0{'}$$
$$A\text{.................}SO\text{..............................}B$$
$$X_1 \quad\quad t_1, t_2 \quad\quad X_2$$

Clearly SO does not <u>see</u> (observe) the flashes simultaneously (because the speed of light is the same in all directions and frames and is not infinite), even though in reality the flashes were emitted simultaneously. However, SO can determine by simple classical calculations whether they were indeed simultaneous or not if SO's position between A and B is known (and the speed of light c is known).

Thus:
$x_1 = c(t_1 - t_0), x_2 = c(t_2 - t_0{'})$
i.e., $x_2 - x_1 = c(t_2 - t_0{'}) - c(t_1 - t_0) = c(t_2 - t_1) - c(t_0{'} - t_0)$

therefore, $t_0' - t_0 = (t_2 - t_1) - (x_2 - x_1)/c$

Thus, if SO determines that $t_2 - t_1 = (x_2 - x_1)/c$, all known or measured values, then $t_0' = t_0$ and the flashes were simultaneous (i.e., $x_2 = ct_2$ and $x_1 = ct_1$). Thus, as in example 1 above, <u>simultaneity occurred and could be confirmed by the observer.</u>

3. A more complicated situation (SO is not situated in a straight line with A and B)

Figure 1

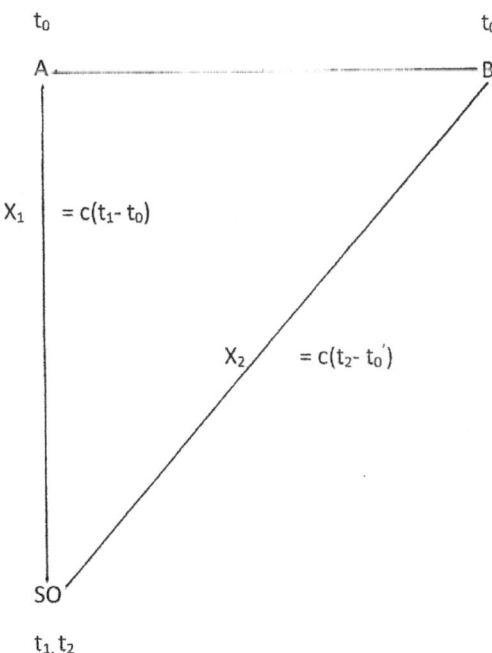

In this situation, as in many more complex situations, the issue becomes more complicated because of the increasing number of unknowns, but if the observer knows the positions of A and B and his position relative to A and B, and he measures the time of arrival of the flashes, he can determine (again using simple mathematics) whether the flashes were simultaneous in reality. The calculation is exactly as in example 2 above and if SO determines that $t_2 - t_1 = (x_2 - x_1)/c$ (i.e., $x_2 = ct_2$ and $x_1 = ct_1$), then again $t_0' = t_0$ and the flashes were simultaneous.

Now let's look at the situation for a moving observer, MO, moving at velocity v compared to the stationary observer, SO:

Figure 2

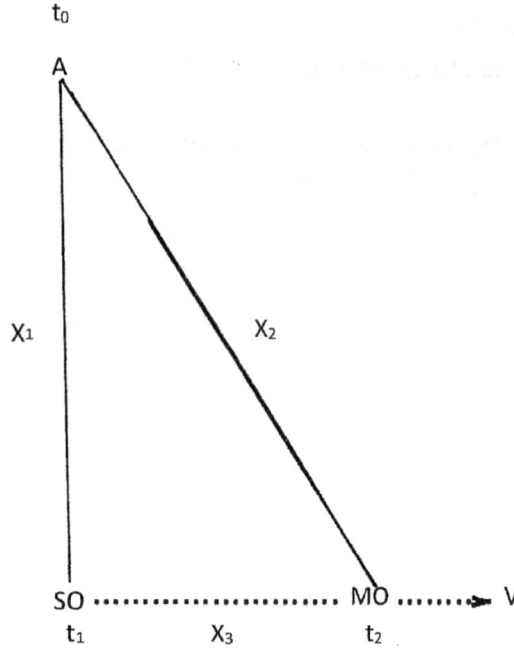

A flash is emitted from site A at t_0. The flash is received at stationary site SO at t_1 and the flash is received at moving site MO (moving with velocity v relative to SO and passing SO at time t_0) at t_2. We can demonstrate the following:

$X_1 = c(t_1 - t_0)$
$X_2 = c(t_2 - t_0)$
$X_3 = v(t_2 - t_0)$

$X_2^2 = X_1^2 + X_3^2$

Therefore,

$c^2(t_2 - t_0)^2 = c^2(t_1 - t_0)^2 + v^2(t_2 - t_0)^2$
i.e., $(t_2 - t_0)^2 = (t_1 - t_0)^2 + v^2/c^2(t_2 - t_0)^2$
i.e., $(t_2 - t_0)^2 (1 - v^2/c^2) = (t_1 - t_0)^2$ or $(t_2 - t_0)^2 = (t_1 - t_0)^2/(1 - v^2/c^2)$

Let $t_0 = 0$, then:

$t_2^2 = t_1^2/(1 - v^2/c^2)$ or $t_2 = t_1/[\text{square root of } (1 - v^2/c^2)]$

which is of course the essence of the Lorentz Transformation (providing the lambda factor 1/[square root of $(1 - v^2/c^2)$]), the basis of Special Relativity, that gives time (t_2) as observed from a moving reference MO, moving at velocity v relative to a stationary observer SO, compared to the time (t_1) observed from the stationary reference SO, and which we have determined using nothing but logic and simple mathematics, compatible with classical mechanics. Note that t_2 is larger than t_1 because the velocity of light c is not infinite and takes longer to reach MO than SO. Note further that if c were infinite – that is, if information transfer were instantaneous – there would be no difference, that is, t_2 would be equal to t_1.

In Appendix 4 I further demonstrate that it is possible to corroborate the reality of simultaneity by means of a simple classical-mechanical-compatible calculation, even from a moving reference system that throws an observational error into the mix. Simultaneity is a reality, but its demonstration may require a simple correction for the observer's point of view, which is what the Special Theory of Relativity provides. Thus, the Lorentz Transformation and its successor, the Special Theory of Relativity, merely provide corrections to match observation, required because the velocity of light c is not infinite, as discussed further below (see also Appendix 7).

Notice again, in the simplified Lorentz Transformation derived above, that if the velocity of light c were infinite, the correction factor for a reference body moving at velocity v would be [1/the square root of $(1 - v^2/\text{infinity})$] = 1, that is, there would be no correction.

Note further that if the velocity v were to approach the velocity of light c, the correction would approach infinity ([1/the square root of $(1 - 1)$] = infinity). Some, including Einstein himself, have interpreted this as evidence that the velocity of light c cannot be exceeded. While it may be true that the velocity of light c cannot be exceeded, that is not the correct interpretation of this calculation. <u>The correct interpretation is simply that observation would be impossible at velocities equal to or exceeding that of light (which should of course be quite obvious)</u>.

Furthermore, what we have shown is that the Special Theory of Relativity is not a revolutionary introduction of a correct new law of reality, but rather is a mere correction of the basic classical law for certain extreme situations, where correcting for observational error is required because the velocity of light is less than infinite. <u>The theory has been so successful because it corrects for observational error – it matches observation – it describes the "observer's world"</u>.

Let's review Einstein's conclusions (9, pp. 30-31) from his analysis of his example regarding the relativity of simultaneity. He says: *"Events which are simultaneous with reference to the embankment are not simultaneous with respect to the train, and vice versa (relativity of simultaneity). Every reference body (co-ordinate system) has its own particular time; unless we are told the reference-body to which the statement of time refers, there is no meaning in a statement of the time of an event."*

This is partly true but would be more correctly stated thus: <u>"Events which are **observed** to be simultaneous with reference to the embankment do not appear to be simultaneous as **observed** from the train. Every reference body (coordinate system) has its own observation of the relative timing of events; unless we are told the reference-body to which a statement of observed time refers, we cannot assess its relationship to the timing we observe from another reference system."</u> Or put another way: <u>Simultaneous events may not be observed to be simultaneous from all reference locations depending on the position and motion of the observer relative to the location(s) of the events. If light must travel a different distance from one event to a location than from the other event to that location, then the events will not appear to be simultaneous from that location, even if they occurred simultaneously in reality.</u> One might then ask: "What are they simultaneous relative to?" The answer, of course, is to each other. If the speed of light were infinite, this would be obvious to all observers.

I must emphasize that every reference frame (coordinate system) recognizes real time in its own frame, but its perception of time in another reference frame may be distorted because of the limitations on information transfer due to the finite speed of light. The Lorentz Transformation, however, allows calculation of the time that will be observed from one reference frame to another. This does not create "special" or "preferred" reference frames because the same is true for all frames – an observer measures time accurately in the observer's own frame, but may perceive time in another frame inaccurately, and what will be observed can be calculated with the Lorentz Transformation. This is equally true for all frames and is not a violation of the first postulate of relativity. The equations of the Special Theory of Relativity (essentially the Lorentz Transformation) provide the required correction to the error in observation caused by the fact that information transfer is not instantaneous (because the speed of light c is not infinite). Thus, the Special Theory of Relativity allows the calculation (prediction) to match what is observed – it reproduces the **"observed world".**

2C   MISLEADING "REVOLUTIONARY" NEW CONCEPTS OF TIME AND SPACE

According to Brian Greene in "The Universe and the Bucket" chapter in *The Fabric of the Cosmos* (4, p. 23 and see Appendix 5), Sir Isaac Newton stated that *"Absolute space, in its own nature, without reference to anything external, remains similar and unmovable."* Apparently, Newton felt that *"Space itself provides the true frame of reference for defining motion"* and *"Absolute space has the final word on what it means to move."*

However, Greene also relates that Leibniz, a contemporary of Newton, *"firmly believed that space does not exist in any conventional sense"* and that *"Without the objects in space, space itself has no independent meaning or existence."* Newton, however, raised further argument and evidence and, apparently, Leibniz was forced to admit, *"I grant there is a difference between absolute true motion of a body and a mere relative change of its situation with respect to another body."* **Amen to this!**

But Greene is firmly on the other side of this argument and proceeds to raise questions about the existence of absolute space and its role as a benchmark for motion, e.g., *"After all, if absolute space really exists it should provide a benchmark for all motion, not just accelerated motion. If absolute space really exists, why doesn't it provide a way of identifying where we are located in an absolute sense, one that does not need to use our position relative to other material objects as a reference point?"* (4, p. 32-33) **This is not a reasonable objection.**

That "absolute space" does not provide a way of identifying where we are located in an absolute sense does not demonstrate that it does not exist. I will deal with why this is so as I proceed (see also Appendices 3 and 5). Furthermore, Greene and many others claim that absolute space doesn't exist, then support the claim that the curvature of this nonexistent space moves matter and explains gravity and that the expansion of this nonexistent entity is carrying the galaxies along with it - **yikes!**

My intuition tells me that space must be infinite and eternal (and, by this, I mean absolute space itself, not the matter, energy or other fields that may occasionally appear in space). There is no other possibility. And my intuition also tells me that time must be infinite and eternal. There is no other possibility. This intuition is the result of many decades of experience, i.e., many decades of observation. So, this intuitive understanding is the result of inductive logic based on countless observations over many decades of time. Inductive reasoning can be compelling but can nevertheless be considered insufficient, particularly if the observation is historical rather than planned, controlled and conclusive.

So how can we introduce some compelling deductive reasoning to this issue? Space, if we get away from the quantum nonsense, is, in actuality, nothing – that is, the absence of something or anything. Right? Think about it for a minute! And time, like space, just is, and must be (i.e., can't not be). They both, in a sense, represent nothing, i.e., empty vessels into which something can be put – and was put. Matter and energy were "put" into space and events/history are "put" into time. And the something "put" into space eventually became us. So, if we try to replace space with nothing what do we get – nothing of course, which is empty space! Similarly, there is no way we can get rid of time – it's with us as long as we are here, and it's with space for eternity – even though it's meaningless (this does not mean nonexistent) without conscious intelligent life to recognize it and ponder it (see note at the end of this section).

There is no real absolute time, because there is no origin, i.e., no absolute t = 0 (because time is infinite and eternal). We see and feel real changes in time. We can accurately measure intervals of time, i.e., the passage of time, but no specific absolute time. The reason for this is obvious – there is no point t = 0 from which to determine it. It is extremely important to understand and accept this (see Appendix 3). The same is true of space. There is no absolute

$x = 0$, $y = 0$, $z = 0$. If the Big Bang Theory were true there might be an origin with absolute zeroes, but this is unlikely, as I will discuss later.

Physicists lament that their laws/equations don't incorporate time or recognize the unidirectionality of time. This is because time (as well as space) has no physical substance, no measurable properties. For space and time, physicists substitute measurements of length and duration, using physical instruments that are subject to their physical laws. Therefore, the quantum- or relativity-based phenomena they describe (e.g., time dilation, length contraction and the curvature or expansion of space) relate to the substitute measurements they make with their physical tools, not to actual time or space. So indeed, I can agree that *"space and time top the list of age-old scientific constructs that are being fantastically revised"* (4, p. 6), but unfortunately fantastically and incorrectly!

In Chapter 12 of his treatise on relativity (9), Einstein introduces us to the concepts of "length contraction" and "time dilation" through a discussion of the behaviour of measuring-rods and clocks in motion. He says, *"I place a metre-rod in the x'-axis of K' in such a manner that one end (the beginning) coincides with the point x' = 0, whilst the other end (the end of the rod) coincides with the point x' = 1."* Then he asks, *"What is the length of the metre-rod relatively to the system K? In order to learn this, we need only ask where the beginning of the rod and the end of the rod lie with respect to K at a particular time t of the system K. By means of the first equation of the Lorentz transformation the values of these two points at the time t = 0 can be shown to be*

x (beginning of rod) = 0 x the square root of [1 - v²/c²] = 0
x(end of rod) = 1 x the square root of [1 - (v²/c²]

the distance between the points being the square root of [1 - v²/c²]. But the metre-rod is moving with velocity v relative to K. It therefore follows that the length of a rigid metre-rod moving in the direction of its length with a velocity v is the square root of [1 - (v²/c²] of a metre. The rigid rod is thus shorter when in motion than when at rest, and the more quickly it is moving, the shorter is the rod."

This has been called "length contraction". But this is not what it seems; the rod does not really shorten (I will prove this further on) – the Lorentz Transformation merely corrects observation for the error introduced by the fact that the speed of light is finite. The metre-rod is merely perceived from K to be shorter than it actually is in K'. This correction is negligible at normal velocities and only becomes meaningful at velocities such that v^2/c^2 becomes significantly greater than 0 (i.e., v approaches c, the velocity of light).

Einstein continues, *"For the velocity v = c we should have the square root of [1 - v²/c²] = 0, and for still greater velocities the square-root becomes imaginary. From this we conclude that in the theory of relativity the velocity c plays the part of a limiting velocity, which can neither be reached nor exceeded by any real body."* Again, this is not the proper conclusion – it may be true (based on other considerations than the Lorentz Transformation) that the velocity c cannot be reached or exceeded by any real body, but, as I have noted earlier, all the Lorentz Transformation says is that observation becomes impossible when "v reaches or exceeds c".

Einstein continues further, *"If we had based our considerations on the Galilean transformation we should not have obtained a contraction of the rod as a consequence of its motion."* Exactly, because the rod did not actually contract, it merely appeared to from the reference frame moving at velocity v relative to the rod, and the Galilean Transformation, unlike the Lorentz Transformation, does not provide a correction for the error in observation caused by the finite speed of light. The Galilean system provides reality in a system at rest, and at normal velocities far below the speed of light. The structure of the "Lorentz Transformation" (the basis of the theory of relativity's correction of observation) is easily calculated, however, as I showed earlier in this chapter when dealing with the concept of simultaneity (see also Appendix 7).

Einstein then moves on to deal with time: *"Let us now consider a seconds-clock which is permanently situated at the origin (x' = 0) of K'. t' = 0 and t' = 1 are two successive clicks of this clock. The first and fourth equations of the Lorentz transformation give for these two ticks:*

$$t = 0$$
and $$t = 1/\sqrt{1 - v^2/c^2}$$

As judged from K, the clock is moving with the velocity v; as judged from this reference body, the time which elapses between two strokes of the clock is not one second, but $1/\sqrt{1 - v^2/c^2}$ seconds, i.e. a somewhat larger time. As a consequence of its motion, the clock goes more slowly than when at rest."

This has been called time dilation. This has to do with the action, or perception of the action, of the clock, but has nothing to do with the dimension of time. Again, the clock does not actually move slower but is erroneously perceived/observed to from a reference frame in relative motion at velocity v. The Lorentz Transformation merely provides the correction, required due to the finite speed of light, to calculate what will be perceived.

I will now provide a very convincing proof of the interpretation I have provided above that length contraction doesn't actually occur, in spite of Einstein's suggestion and the acceptance of this suggestion by a majority of folks. A slightly more complicated but similar exercise can be done to show that time dilation is not real but merely an observational error. In Chapter 3, I will discuss the problems with the various experimental confirmations of the Theories of Relativity and quantum mechanics (or at least the interpretations thereof) that have been reported.

Figure 3

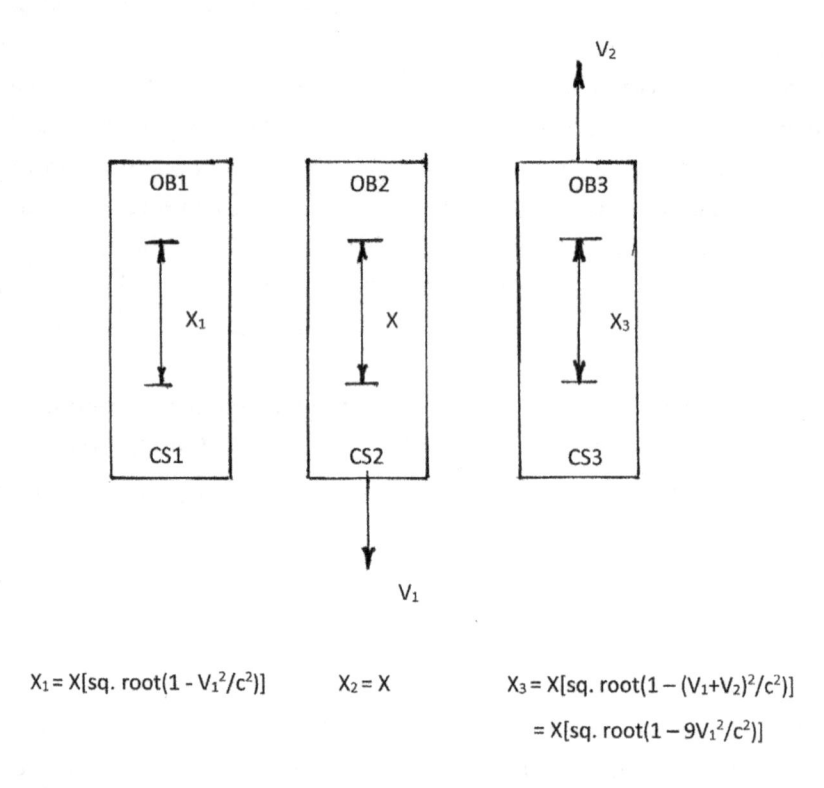

$X_1 = X[\text{sq. root}(1 - V_1^2/c^2)]$ $X_2 = X$ $X_3 = X[\text{sq. root}(1 - (V_1+V_2)^2/c^2)]$
$\qquad\qquad\qquad\qquad\qquad\qquad\qquad\qquad\qquad\quad = X[\text{sq. root}(1 - 9V_1^2/c^2)]$

I show here three different observers in three different coordinate systems: CS1, CS2 and CS3. CS1 is at rest, CS2 is moving south at velocity v_1 relative to CS1, and CS3 is moving north at velocity $v_2 = 2v_1$ relative to CS1. The observer in CS2 has a rod moving with him at velocity v_1 relative to CS1 that he measures to have the length x. Now let's apply the methodology Einstein used to demonstrate what he called **length contraction** as reviewed on the previous page. According to the observer in CS1 the length of the rod appears to be x [square root of $(1 - v_1^2/c^2)$] as dictated by relativity and the Lorentz Transformation. And, in contradiction to this, according to the observer in CS3, the length of the rod appears to be x[square root of $(1 - (v_1+v_2)^2/c^2)$] = x[square root of $(1 - 9v_1^2/c^2)$], again as dictated by relativity and the Lorentz Transformation. So, to which length did the rod in CS2 contract – the one observed from CS1 or the one from CS3? Clearly it can't be both.

In fact, we could examine an infinite number of coordinate systems with different relative velocities and get an infinite number of contracted lengths for x. Note further that by the theory of relativity, the length of the rod would appear to change even if the rod were stationary and the observers were moving relative to the rod. Thus, the rod appears to change in length even though it is not moving, but the observer is. Thus, it is obvious that we are dealing with observation issues, not reality. It is generally understood that the

Special Theory of Relativity was developed based on two assumptions or postulates. In his book on the subject (9), Einstein phrased these postulates thusly:

1. The principle of relativity: if relative to K, K' is a uniformly moving coordinate system devoid of rotation, then natural phenomena run their course with respect to K' according to exactly the same general laws as with respect to K.
2. The velocity of transmission of light in vacuo has to be considered equal to a constant c.

These postulates have since been expressed as follows:

1. The laws of physics are the same in all inertial frames of reference.
2. The speed of light in vacuum has the same value c in all inertial frames of reference.

Of course, the second "law" is not a postulate but a demonstrated fact and the first "law" is demonstrably obvious – furthermore, a natural law must be a natural law even in a non-inertial system, although the equations expressing these laws may need modification to correct for certain special circumstances. Deriving the Special Theory of Relativity in this way from these postulates gave the impression that a new law of nature was being derived, when in reality this was just a way to derive the Lorentz Transformation while obfuscating the fact that it was just a correction required for observation because the speed of light c is not infinite (see Appendix 7).

In Appendix One to his book (9), Einstein provided a "Simple derivation of the Lorentz Transformation", which initially bothered me with its obfuscation. Apparently, others had a similar reaction. RJ Anderton (20) reviewed this derivation and concluded that *"Einstein made numerous math mistakes, his special relativity is a collection of math mistakes, and modern physics still uses that collection of mistakes to teach physics students adding extra mistakes as it goes. So, the issue is to try to emphasize one specific mistake among his many which makes it clear that existing SR is a farce needing revision."* This seems a bit harsh to me – what I have shown is the following: what the mathematical exercise does, in actuality, is simply derive the Lorentz Transformation, which provides a convenient blind formula for physicists to use to make certain calculations.

Earlier in this chapter I demonstrated that, contrary to Einstein's suggestion, simultaneity occurs in actuality (reality), but there is relativity in the **observation** of reality. I have provided in this Chapter (see Fig. 2 in section 2b) a very simple derivation of the lambda factor of the Lorentz Transformation that shows that it simply corrects the calculation to what will be observed, required due to the fact that information transfer is not instantaneous because the constant speed of light c is not infinite (see also Appendix 7).

Back to length contraction: clearly length contraction is not real – nothing contracted. The stationary length is x, and this is the "real" length – the rod doesn't contract, it just appears to do so from different points of view – and the relativistic Lorentz Transformation merely calculates the observer's distorted measurements due to the finite speed of light. And this explains why physicists always get the correct answer when they apply these equations – because they calculate what will be observed under the existing conditions, i.e., in the relevant coordinate system (unless they use the wrong coordinate system; see e.g. ref 1C).

So here is how I see the universe and the concept of creation that I will expand upon in the last two chapters. Initially there were only space and time. These can be neither created nor eliminated, they are both infinite and eternal. Therefore, there is no origin, no zero time, and there is no uniqueness to different points in space. This is an important basis of Einstein's "relativity" concept that has been somewhat misinterpreted, leading to some of the quantum absurdities I am addressing in this treatise. Unless we establish an artificial origin, we can't distinguish between points in space. We can talk about periods of space and time, dx and dt, based on artificial frames of reference, but these would have no absolute meaning with respect to the realms of infinite and eternal space and time. In fact, these concepts, indeed nothing, had "meaning" until the "creation" of matter. With this advent, there was still no absolute origin to space and time, but now there was a basis for coordinate systems to give meaning to dx and dt. Eventually consciousness evolved, which was able to observe, followed by the evolution of intelligence, which was able to deduce. <u>One of the problems with quantum mechanics and the Theories of Relativity is that they are based on an infinite number of possible origins or frames of reference (coordinate systems). This leads to multiple solutions of the equations, suggesting to some, probably erroneously, the existence of multiple dimensions and multiple universes that we see no evidence of.</u> I will expand on this important point elsewhere in this treatise.

<u>In fact, the theory of relativity is the problem – it removed the absolute reference system (i.e., space) and replaced it with an infinite number of artificial or surrogate systems.</u> This allows physicists to calculate the correct observed values from their selected reference frame, but prevents the equations from calculating anything back to a beginning (i.e., origin of the universe). Furthermore, this approach provides absurdities outside a limited frame of reference.

Note: In pure empty space, time is totally meaningless (as is "position"), and some might say it does not exist. However, once matter appears in space, time gains some meaning (as does "position"), but it requires consciousness to provide an observer to recognize that meaning and provide a measure of "duration" – not necessarily a measure of time, but an observational measure of "duration" (which requires something observable, i.e., matter). I must state again: time and space are real, absolute, infinite and eternal, but have no inherent physical properties we can measure, other than the surrogate measures of distance and duration.

I will now discuss the problems with the interpretation of the General Theory of Relativity.

2D THE PROBLEMS WITH THE GENERAL THEORY OF RELATIVITY

In his article "What Einstein Got Wrong" in Scientific American (1c), Lawrence Krauss states: *"General Relativity tells us that nature is independent of the particular way that scientists choose to define coordinates in space;"* but *"many seemingly bizarre results that come out of solving relativity's equations are now understood as mere artifacts of using the wrong coordinate system."* Doesn't seem to make much sense but this is explained by what I have been presenting.

Let's take a closer look. Consider the first quoted sentence – that aspect of relativity (the first postulate, actually), as I have already mentioned, seems to me not to be some brilliant insight or new law but to be totally intuitively obvious (it wouldn't make sense any other way). However, while nature and its physical laws should not be affected by our choice of coordinate system, the specific measurement made during an **observation** could require a correction for the coordinate system chosen. This would explain the apparent contradiction Krauss identified, and indeed, I have shown earlier in this chapter that that is what the Special Theory of Relativity in fact does – it doesn't provide a new law of nature but provides a correction of Newton's laws for observations made under certain special circumstances from a specific coordinate system. This correction applies to certain fast-moving objects or coordinate systems, required because the speed of light is not infinite (i.e., information transfer is not instantaneous).

In the same issue of Scientific American (1d), Walter Isaacson discusses Einstein's "Equivalence Principle" – the idea that gravity and acceleration are equivalent, i.e., that there is no way to make a distinction between the effects of gravity and the effects of being accelerated. He describes Einstein's "chamber thought experiment" wherein a chamber is being accelerated upward and a light beam is shone in through a pinhole in one wall. It is claimed that if you could plot the beam's trajectory it would be curved because of the upward acceleration. Isaacson states that Einstein's goal was to find mathematical equations describing how a gravitational field acts on matter, telling it how to move, and how matter generates gravitational fields in spacetime, telling spacetime how to curve. (Wow, kind of reminds one of that old saw, "Which came first, the chicken or the egg?") And he relates how Einstein asked an old classmate (Marcel Grossman) to help him with the complicated mathematics that might describe a "curved and warped four-dimensional spacetime".

And herein, I believe, lies the problem with the interpretation of Einstein's General Theory of Relativity that I will now try to clarify. One aspect of the problem is that the equations were apparently developed for spacetime, but spacetime and space are not identical. Spacetime is an artificial construct that does not have an actual physical place in reality; space, on the other hand, is real. Physics has evolved to get these two concepts mixed up, with troubling consequences. I will explain this and then explain my correction to the interpretation of the General Theory of Relativity.

Greene (4, p. 59-61) says: "... *Newton definitely got it wrong* .... *Absolute space does not exist. Absolute time does not exist. But according to special relativity, absolute spacetime does exist.*" On p. 61, after discussing three astronauts (one moving without acceleration, one flying repeatedly in a circle and one accelerating into deep space), he claims: "... *we learn that geometrical shapes of trajectories in spacetime provide the absolute standard that determines whether something is accelerating. Spacetime, not space alone, provides the benchmark. ... special relativity tells us that spacetime itself is the ultimate arbiter of accelerated motion. ... special relativity showed once again that the arena of reality – viewed as spacetime, not as space – is enough of a something to provide the ultimate benchmark for motion.*" (I think Greene meant to say General rather than Special Relativity – but that's not the worst of the issues here).

This is all very misleading but opened my eyes to one arena where the interpretation of the Theories of Relativity has gone wrong. First, we are told space is not real and does not exist and time is not real and does not exist, then we are told that combining these two nonexistent entities as spacetime provides the ultimate benchmark for motion. This is disturbing. Then Greene combines a time axis with a space axis to get a curved line. Nothing surprising there, but let's explore this a little further. A point in real 3D space is defined by three coordinate measures: x, y and z, provided from three perpendicular coordinate axes. Such a point, if in motion, is moving along an independent time axis t and x, y and z may change with t. If two of the space coordinates are zero, or constant, the point will be moving in a straight line along the remaining axis, say x, and if an acceleration (force) is applied along this axis, the point will simply move faster along this same straight line (Fig. 4).

Alternatively, the point may be moving along with t in a straight line defined by non-zero coordinates x, y and z. If an acceleration is applied along this same straight line, again the point will simply move faster along that straight line (Fig. 5). If an acceleration (i.e., a force, such as gravity) is applied in a direction different from the original straight-line trajectory in space, then the point will trace a curved trajectory in space (Fig. 6). The trajectory in **space** is the ultimate arbiter of motion and it need not necessarily be curved. Only if you plot the change in position in space per unit time versus time will you always get a curved line (trajectory) for accelerated motion. Space and time are the real independent absolutes and "spacetime" is an artificial construct that sounds reasonable since both space and time are involved.

This is all encompassed by Newton's laws and no corrections are required unless the "motion" occurs at speeds such that v^2/c^2 becomes significantly greater than zero. Because both space and time are involved in this, physicists have defined a convenient entity, "spacetime", which can be useful, but not if used to denigrate space and time. Space and time are independent dimensions that cannot directly affect each other. The position of an object in space can change with time – but this is not due to any action of time on space or the object.

Newton was not wrong, but the quantum enthusiasts are determined to make this claim and that may be part of the reason why relativity and quantum mechanics are incompatible and have gone awry.

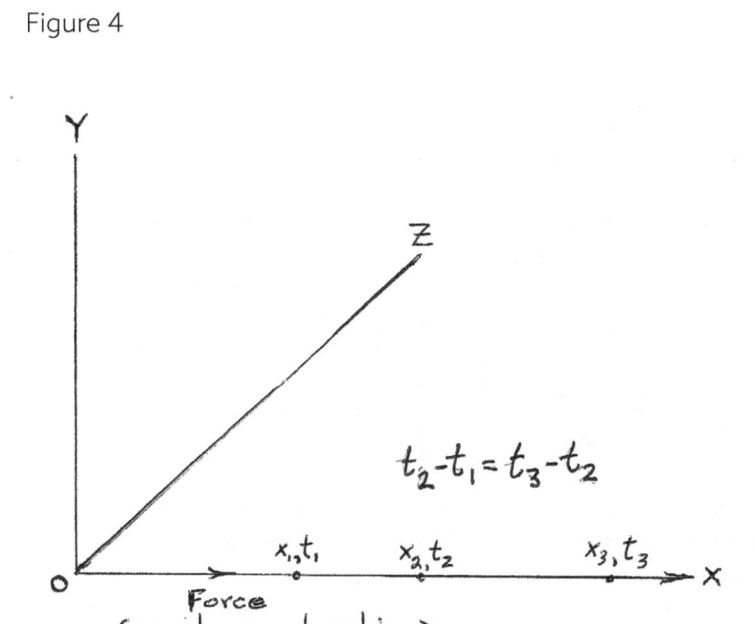

Figure 4

$$t_2 - t_1 = t_3 - t_2$$

Force (provides acceleration)

A point x is moving with time t in a straight line along the x axis. If an acceleration (Force) is applied along this axis, the point will simply move faster along this same straight line as shown.

Figure 5

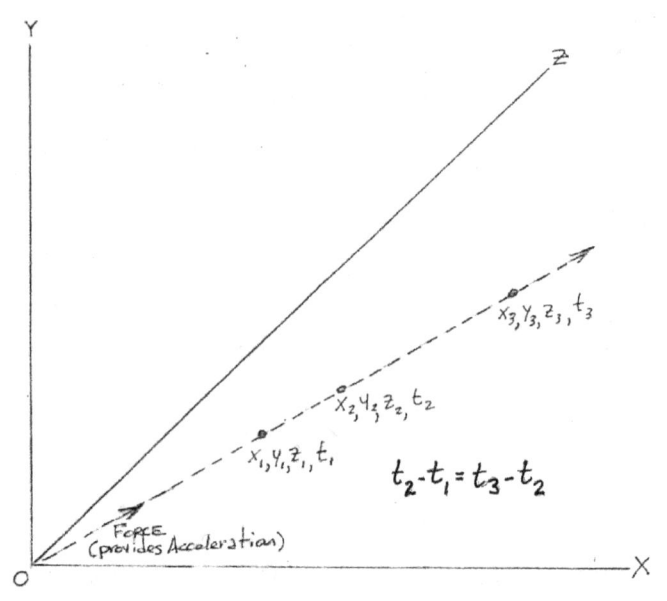

A point is moving along with time t in a straight line defined by non-zero coordinates x, y, z. If an acceleration (Force) is applied along this same straight line, again, that point will simply move faster along that same straight line.

Figure 6

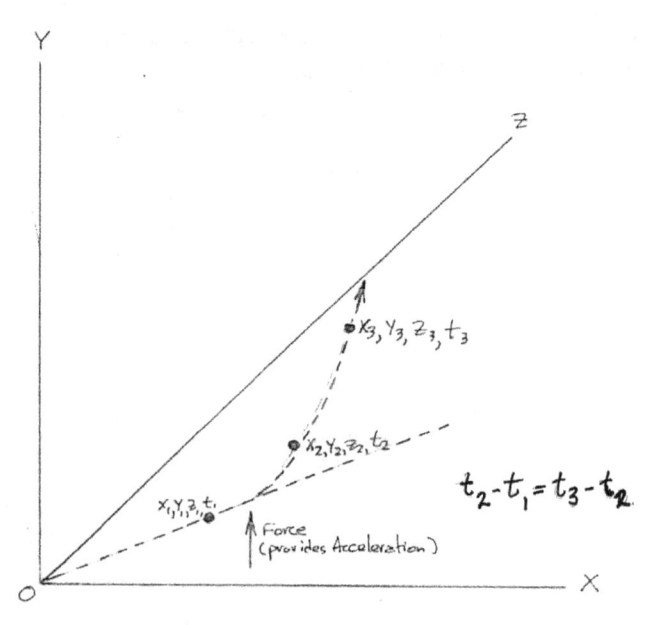

On the other hand, if a point is moving along with time t in a straight line defined by non-zero coordinates x, y, z and an acceleration (Force) is applied in a direction different from the original straight line trajectory in space, then the point will trace a curved trajectory in space.

Now let's take a closer look at the General Theory of Relativity and its accepted interpretation. The General Theory is hailed as a revolutionary new theory that topples Newton and explains gravity. Really, and how does it do this? Well, by suggesting that gravity in fact curves and warps space such that matter follows this curve, just like a ball will fall down the curve of the surface of a trampoline indented by a heavy weight such as a big rock or an athlete jumping on the trampoline. Really!!! Well, isn't that interesting – and here I always thought that the explanation for a ball rolling down a curve required gravity as an explanation! So, the effect of gravity is explained by the effect of a curve that is explained by the effect of gravity! But that is circular reasoning, which is no reasoning at all!

But wait a minute – that isn't the only problem with this problematic theory! Even if the curvature of space could explain gravity (which it can't), how can anything curve space? How can you curve nothing? That's impossible! First, we're told that space doesn't really exist, and it's just a concept we use to explain the separation between objects of matter – then we are told that the curvature of space explains gravity. Yikes, my head hurts. OK, back to the facts. Space is real but it consists of nothing that can be curved (forget things in space that can be curved such as matter, energy and other fields – these are not "space"). And even if it could be curved, that wouldn't explain gravity! I discuss more problems with this theory in Section 3e (e.g., the creation of a white dwarf or a neutron star or the "big crunch" while standing on a neutron star). **But the equations work, so if the interpretation is wrong, what is the correct explanation?**

Here is my explanation. But let me digress for a moment to explain my explanation. The equations of General Relativity are very complex, and I have not analyzed and broken them down to prove what I am about to suggest – not many, if any, could do that. I have merely scoured my brain's database to come up with something that might explain the effectiveness of these equations more accurately than the accepted theory (i.e., warping of space) does. I suggest that the equations work because they correctly describe the trajectory that the object in question follows in space. Gravity doesn't somehow curve space, which then somehow makes the object follow this curve (as I explained above) – rather gravity provides a force that causes the object to follow a curved trajectory through space, and the equations reproduce this trajectory. (This curved trajectory in space takes time to complete – so both space and time are involved, making spacetime a convenient artificial construct to refer to, but there is no "fabric of spacetime".) The curved trajectory in space caused by a force (gravity) on a moving body is completely encompassed by Newton's "classical mechanics" and only requires a correction with so-called "relativistic" concepts if the motion approaches a significant proportion of the speed of light. The correction is for what will be observed and in the case of Special Relativity only becomes meaningful when v^2/c^2 becomes significantly greater than zero. It's a similar situation but a little more complicated for General Relativity.

I showed previously how the Special Theory of Relativity applies the Lorentz Transformation, in cases of constant velocity, to correct errors in observation caused by the non-infinite speed of light, i.e., the non-instantaneous transfer of information. This error, and its correction by the Lorentz Transformation, depends on the velocity v of the observed object relative to the observational frame of reference, and is only significant if the ratio v^2/c^2 is significantly greater than zero. The application of this correction to an accelerated object is much more complicated because in this case v is changing with time.

In fact, this was so complicated that it took Einstein ten more years, with the help of his mathematician friend Marcel Grossman, to develop the necessary equations. These equations dealt with the complications inherent in the application of the transformations required by the changing velocities, as well as the iterative complications engendered because a gravitational field, generated by a mass, has energy. Thus, by Einstein's famous equation $E = mc^2$, it has mass, and thus the gravitational field, generates a further gravitational field, and so on ad infinitum (similar to the reflections of a mirror in a mirror). Of course, eventually the gravitational fields generated in this progression become functionally negligible, but nevertheless this phenomenon had to be accounted for.

According to Cole (11, p. 26), "*What he had to find was a set of laws* [I would rephrase this as "a set of equations"] *that could deal with any form of accelerated motion and any form of gravitational effect.*" Working with his friend Marcel Grossman, who helped him learn and apply sophisticated mathematical techniques such as tensor analysis and Riemannian geometry, Einstein managed to "*invent a formalism that was truly general enough to describe all possible states of motion.*" Quoting Cole again: "*Understanding the technicalities of the general theory of relativity is a truly daunting task, and calculating anything useful using the full theory is beyond all but the most dedicated specialists. While the application of Newton's theory of gravity requires one equation to be solved, Einstein's theory represents 10 independent equations which are all non-linear. .... This nonlinearity leads to unmanageable mathematical complexity when it comes to solving equations. But the crucial aspect of this theory is that it relates the properties and distribution of matter to the curvature of space. ... The idea that space could be warped is so difficult to grasp that even physicists don't really try to visualize such a thing.*"

I go one step further here and claim that the curving or warping of space is not just "difficult to grasp", but cannot and does not occur. In fact, what Einstein and Grossman's complicated mathematical equations probably do is take into account all the possible factors in detail, such as complicated transformations like the iterative application of the Lorentz Transformation, to correct for accelerated motion and iterative gravitational effects. (That is, fields creating fields. Because of the equivalence between mass and energy through Einstein's equation $E = mc^2$, all forms of energy gravitate – the gravitational field produced by a body is itself a form of energy, and therefore also gravitates.)

These new equations by Einstein and Grossman thus provided a more accurate solution than the simple equation of Newton, but in the end what they describe is the curved trajectory over time of the accelerated object in space – there is no curvature of space. I suggest that Einstein's equations, which are very complicated extensions of Newton's laws, apply to the motion of matter in space, not to the behaviour of space. I must again emphasize that I have not mathematically proven the explanation that I have given for the meaning of the equations of General Relativity. I have provided the best explanation that one of my knowledge and ability could provide. The one thing I am sure of is that gravity is not explained by a curvature or warping of space. I only hope that my mental effort will inspire someone with greater mathematical expertise to prove me right or wrong.

Before continuing, I must emphasize again that I have immense respect for Albert Einstein – he is in fact my hero. The equations he and Grossman developed were a great accomplishment, but what I most respected was his intuitive discomfort with quantum mechanics. The irony is that some of the careless interpretations of his Theories of Relativity, enhanced to some degree by his own compulsion to promote these theories, have contributed to the enigma that he abhorred and spent so much of his later years trying to resolve.

The bottom line is that the Theories of Relativity and quantum mechanics accurately depict what is observed, or might be observed, and only this. The inability to recognize this is the basis of some of the absurdities that flow from their interpretation. These theories successfully describe the "observer's world", but not necessarily the "real" world. Contributing to these absurdities is the problematic Copenhagen Interpretation of quantum mechanics (this relates to the "observed world", not the "real" world) and the equally problematic suggestion that the curving of space explains gravity.

Note: Nothing exists in the quantum world until it is observed – but everything continues to exist in the real world! Einstein was right. Quantum mechanics works for everything it has been tested for so far, but it is not complete and can produce "absurdities".

2E FURTHER EVIDENCE AGAINST THE CURVATURE OF SPACE AS AN EXPLANATION FOR GRAVITY

I explained above why I believe the curvature of space is impossible and wouldn't explain gravity, even if curvature of space were possible, and have further explained what I believe the equations of General Relativity might really be saying. Here is further evidence that curvature of space is not an explanation for gravity.

i. Neutron Stars

On p. 13 of *Quantum Theory Cannot Hurt You* (10), Marcus Chown says: *"… a way does exist to squeeze the matter of a massive star into the smallest volume possible. The squeezing is done by tremendously strong gravity, and the result is a neutron star. Such an object packs the enormous mass of a body the size of the sun into a volume no bigger than Mount Everest."*

On p. 48 he says: "*A new balance is struck with the inward pull of gravity balanced not by the outward force of the star's hot matter but by the naked force of its electrons. Physicists call it degeneracy pressure. But it's just a fancy term for the resistance of electrons to being squeezed too close together. A star supported against gravity by electron pressure is known as a white dwarf. Little more than the size of earth and occupying about a millionth of the star's former volume, a white dwarf is an enormously dense object. A sugar cube of its matter weighs as much as a car! ... But the electron pressure that prevents white dwarfs from shrinking under their own gravity has its limits. The more massive a star, the stronger its self-gravity* [try to explain this by a curvature of space]. *If the star is massive enough, its gravity will be powerful enough to overcome even the stiff resistance of the star's electrons. ... Eventually the star shrinks so much that its electrons, despite their tremendous aversion to being confined in a small volume, are actually squeezed into the atomic nuclei* [**and try to explain this by a curvature of space**]. *There they react with protons to form neutrons, so the whole star becomes one giant mass of neutrons."* **Obviously, this cannot be explained by a curvature of space.** As I consider this, I recall that Big Bang theorists claim the whole universe was just prior to the Big Bang squeezed into a space less than the size of a proton – one single proton (in fact, much smaller than a proton – a space of zero volume)! This is obviously ludicrous, and I will get back to this in the next two chapters. In the meantime, back to curvature of space explaining gravity.

TRY TO IMAGINE THE PHENOMENA DISCUSSED ABOVE AS DUE TO A CURVATURE OF SPACE! This doesn't make any sense to me, and there are many more examples like this where it is beyond impossible to imagine the curvature of space as the causative factor. It's probably not necessary to describe in detail all the myriad circumstances where gravity is at work, where it is obviously impossible to invoke a curvature of space as the causative factor. Some might try to say "Oh, but it's actually the curvature of spacetime that's responsible," but that doesn't change the conclusion that gravity does not work by curving space. As mentioned previously, "spacetime" represents an artificial construct and plotting position in space versus time can produce a curve in certain circumstances, but there is no curvature of space and can't be.

Let me now try to briefly describe how I think we got to this provocative but impossible concept, after a brief diversion regarding "the force of gravity".

ii. What?? "The Force of Gravity Does Not Exist"??

On p. 113 of *Quantum Theory Cannot Hurt You* (10), discussing the source of atomic or nuclear energy, Chown says, "*... where did all this energy come from? The answer is from gravity. Gravity is a force of attraction between any two massive bodies. In this case* [a simple example he was discussing], *the gravity between earth and the slate pulls them closer together.*"

Reasonable enough, and entirely consistent with Newtonian gravity, although it is puzzling how one can say something like this and still accept the "curvature of space" explanation of gravity. And then in the chapter "The Force of Gravity Does Not Exist", Chown proceeds to defend the odd contention of this title with the following statement on p. 122: "*All massive bodies, once set in motion, have a tendency to keep traveling at constant speed in a straight line.*" No argument here, this is Newton's infallible first law of motion, of course.

But then he proceeds to discuss a car negotiating a curve: "*Because of this property, known as inertia, unrestrained objects inside the car, including a passenger like you, continue to travel in the same direction the car was travelling before it rounded the bend* [this is true, but…]. *The path followed by the car door, however, is a curve* [… at least it's not suggested that the car curved space, and that the curved space is now carrying the car along!]. *It should be no surprise, then, that you find yourself jammed up against a door. But the car door has merely come to meet you in the same way that the floor of the accelerating spacecraft came up to meet the hammer and the feather* [referring to a previous example similar to the one Einstein used in his dissertation on General Relativity]." And Chown concludes, "*There is no force.*"

WHAT! This is not right – there clearly is a force! A force was exerted on the car, by the driver through the intermediation of the steering wheel and its mechanisms, to make its trajectory curve, and the car door, with its direction of motion changed by this force, exerts a force on the "YOU" in this little scenario to change "YOUR" direction of motion. So, we have not been convinced by this little example that "*the force of gravity does not exist*"! Now let me continue with the brief story of how these odd concepts came to be.

iii. The Principle of Equivalence

The concept that gravity is warped space seems to be broadly accepted. I, on the other hand, find this very hard, shall I say impossible, to accept. I will try to explain this further by reviewing the approach Einstein used to develop the concept, using a version published by Peter Cole (11). I will present this version then present my reactions to the analysis and reasoning.

"*Following this* [Einstein's] *lead, we can ask what kind of path light rays follow according to the general theory of relativity. In Euclidean geometry, light travels on straight lines* [Light does not travel in Euclidean geometry – Euclidean geometry is a mathematical construct to describe space and changes in position in space. In fact, light travels in straight lines in space unless a force acts to change that]. *We can take the straightness of light paths to mean essentially the same thing as the flatness of space. In special relativity, light also travels on straight lines, so space is flat in this view of the world too. But remember that the general theory applies to accelerated motion, or motion in the presence of gravitational effects. What happens to light in this case? Let us go back to the thought experiment involving the lift. Instead of a spring with the weight on the end, the lift is now equipped with a laser beam that shines from side to side. The lift is in deep space, far from any sources of gravity. If the lift is stationary, or moving with constant velocity, then the light beam hits the side of the lift exactly opposite to the laser device that produces it. This is the prediction of the special theory of relativity. But now imagine the lift has a rocket which switches on and accelerates it upwards. An observer outside the lift who is at rest sees the lift accelerate away, but if he could see the laser beam from outside it would still be straight. He is not accelerating* [?? The lift is accelerating relative to him, so he is accelerating relative to the lift!] *so the special theory applies to what he sees. On the other hand, a physicist inside the lift notices something strange. In the short time it takes light to travel across, the lift's state of motion has changed (it is accelerated).* [Good grief – the lift is not accelerating relative to the physicist observer inside the lift.] *This means that the point at which the laser beam hits the other wall is slightly below the starting point on the other side. What has happened is that the acceleration has bent the light ray downwards.* [NO, the light beam has not been bent, the wall has simply moved up a bit!]

"*Now remember the case of the spring and the equivalence principle. What happens when there is no acceleration but there is a gravitational field* [Hmm, first we are told that acceleration and gravitation are identical (equivalent), then we are told there is a gravitational field but no acceleration!!!] *is exactly the same as in an accelerated lift. Consider now a lift standing on the earth's surface. The light ray must do exactly the same thing as in the accelerating lift; it bends downward. The conclusion we are led to is that gravity bends light. And if light paths are not straight but bent, then space is not flat but curved.*" NO, this does not follow. The fact that a light path may be "bent" does not dictate that space is curved – light may be deflected, but this doesn't require space to be curved!

Now, look at the figure provided with this little exercise (Fig. 7). This shows what this author proposes. But there seems to be an error here that may have led to the problem I see in the interpretation of the General Theory of Relativity. In Fig. 8, I try to show what really happens.

Figure 7

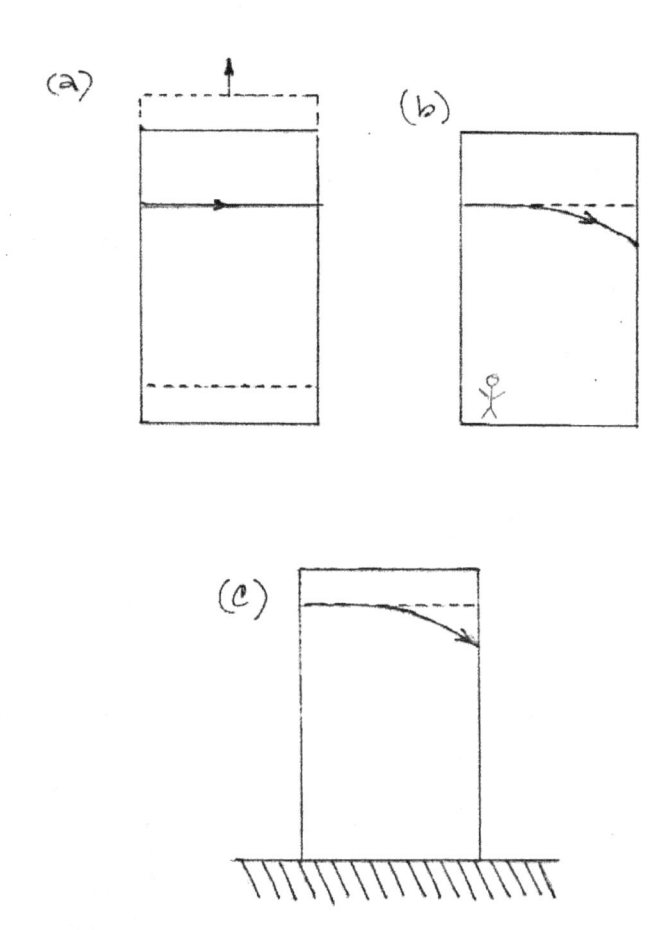

The bending of light. In (a), our lift is accelerating upwards. Viewed from outside, a laser beam follows a straight line. In (b), viewed inside the lift, the light beam appears to curve downwards. The effect in a stationary lift situated in a gravitational field is the same, as we see in (c).

Let's repeat the author's conclusions: "*What has happened is that the acceleration has 'bent' the light ray downwards. … What happens when there is no acceleration but there is a gravitational field, is exactly the same as in an accelerated lift. Consider now a lift standing on the earth's surface. The light ray must do exactly the same thing as in the accelerating lift; it bends downward. The conclusion we are led to is that gravity bends light. <u>And if light paths are not straight but bent, then space is not flat but curved.</u>*" **As I have already stated this last sentence does not logically follow.**

This is apparently based on Einstein's acceptance of the law of the propagation of light (the velocity of light must be constant), therefore if the light beam is "bent" this must be because space is "bent". But I don't believe that this is a necessary or reasonable conclusion. There are ample examples of light "bending" without the need for space to be curved (e.g., passing from one medium to another, air to water or air to glass, etc.). Furthermore, light is bent by gravity because it is energy and has a small equivalent mass according to Einstein's $E = mc^2$.

Let me now try to demonstrate where I think this has gone wrong, possibly contributing to the error in the interpretation of the General Theory of Relativity.

First let me comment on some oddities of the example in Fig. 7. The gravitational field required to accomplish anything measurable resembling Fig. 7(c) would be astronomically immense, as would the acceleration required to produce anything measurable in Fig. 7(b). But I will accept these conditions as simply demonstrative and start my further analysis by providing a slight revision of Fig. 7.

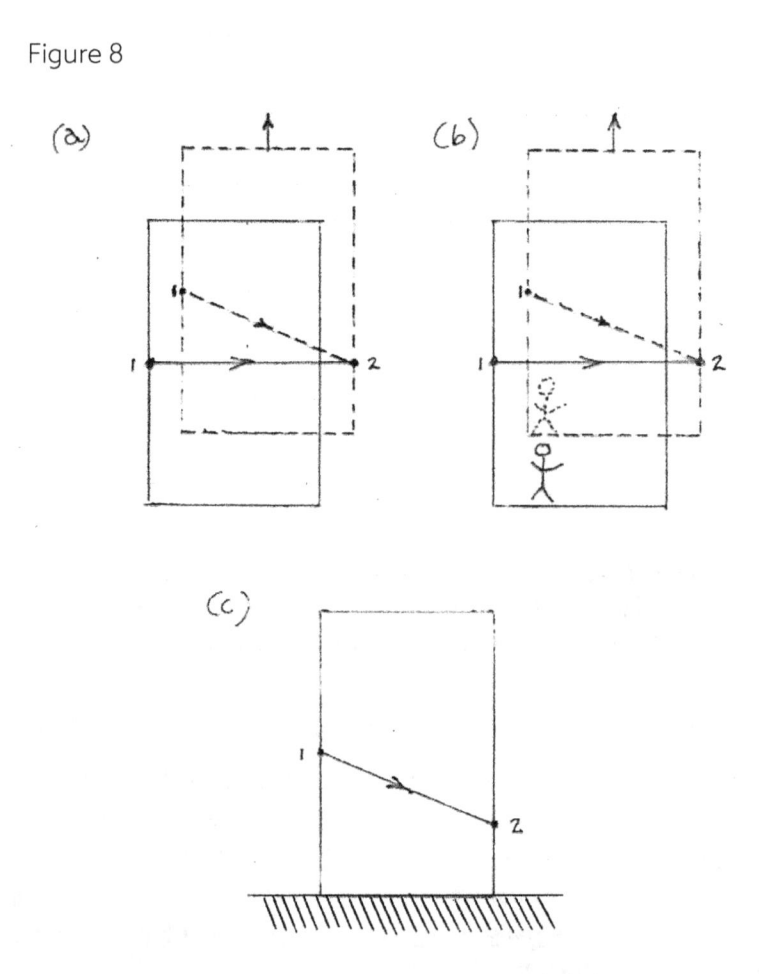

Figure 8

In 8(a) I show a lift being accelerated upwards as in Fig. 7. A laser beam is emitted from point 1 and hits the opposite wall at point 2. This light beam travels in a straight line through space (the only real absolute coordinate system). If we marked points 1 and 2 on the walls and stretched a string between these two points, we would get Fig. 8(a) and 8(c). In Fig. 8(b) everything is identical to 8(a) except there is an "observer" in the lift. Of course, the presence of the "observer", and whatever he observes, is irrelevant to reality, which is identical to what we just described with 8(a) and 8(c). And note further that Cole gets something else completely wrong. In fact, the external observer is accelerating relative to the lift while the physicist inside the lift is not. Concerning Fig. 7(c) and 8(c), the light beam in these situations is deflected because gravity exerts a force on the mass equivalent of the photons.

Note that, in Cole's previous discussion of Fig. 7 (ignoring his error in indicating who is and who is not accelerating relative to the lift), he is attempting to relate not what is real but what is being observed. The conclusion to derive from this little exercise is that no evidence for the curvature of space has been provided and we must take a more careful look at the "Principle of Equivalence".

So, to repeat, the issues with the General Theory may derive from Einstein's reliance on his Principle of Equivalence in developing his theory. He had a eureka moment in thinking about free fall by assuming that objects in free fall do not experience being accelerated downward (towards the earth) but rather weightlessness and no acceleration.

Walter Isaacson provides an account of Einstein's eureka moment in his article "How Einstein Reinvented Reality" (1d): *"The general theory of relativity began with a sudden thought. It was late 1907.... While sitting in his [Einstein's] office in Bern, a thought 'startled' him, he recalled: 'If a person falls freely, he will not feel his own weight.' He would later call it 'the happiest thought in my life.' ... Einstein soon refined his thought experiment so that the falling man was in an enclosed* chamber, *such as an elevator* [lift – as we have already discussed above], *in free fall. In the chamber, he would feel weightless. ... There would be no way for him to tell* [not really, he knows he's not in outer space – if you were in an elevator in which the cable had broken and the elevator was in free fall, you would definitely notice something was dangerously wrong!] – *no experiment he could do to determine* – *if the chamber was falling at an accelerated rate or was floating in a gravity-free region of outer space* [??? there must be a force in the first case and there isn't in the latter case]. *Then Einstein imagined that the man was in the same chamber way out in space, where there was no perceptible gravity, and a constant force was pulling the chamber up at an accelerated rate. He would feel his feet pressed to the floor. If he dropped an object it would fall to the floor at an accelerated rate – just as if he stood on earth. There was no way to make a distinction between the effects of gravity and the effects of being accelerated. Einstein dubbed this 'the equivalence principle'. The local effects of gravity and of acceleration are equivalent. Therefore, they must be the manifestations of the same phenomenon* [this does not necessarily logically follow], *some cosmic field that accounts for both acceleration* [note that in this example it is a force, not a field, that is causing the acceleration] *and gravity* [in fact, gravity is a force field that causes acceleration!]. *It would take another 8 years for Einstein to turn his falling-man thought experiment into the most beautiful theory in the history of physics."*

Unfortunately, there is a problem with this "sudden thought" that became a "beautiful theory". Let's consider this "falling-man thought". Free fall towards earth is not the same as the experience of floating in outer space devoid of all gravitational influence – this is a fallacy. The man may not <u>feel</u> the force on him (observation) because there is nothing pushing back. But he is indeed experiencing a force and being accelerated towards earth – potential energy is converting to kinetic energy that he will become aware of when he reaches earth, and this kinetic energy does work on his body and the ground, and is eventually converted to thermal energy. This same issue holds regarding the "diving airplane" <u>simulation</u> of the gravity-free condition of outer space. This provides the <u>appearance</u> of a gravity-free condition, but only the <u>observer-perceived</u> appearance of same – the "astronaut" is free falling, but so is the airplane. Both are being accelerated earthward at the same rate, giving the astronaut the impression that he/she is floating, but this is just the false appearance of weightless floating. I would rephrase Einstein's principle as the "principle of some similarity" rather than "equivalence". At any rate, Einstein considered the conditions indistinguishable and therefore equivalent, and proceeded for the next eight years to construct his General Theory of Relativity with this conviction.

In Chapter 3, I discuss in some detail the problems with Eddington's 1919 demonstration of the "bending" of starlight as a confirmation of Einstein's General Theory. Note that Newton's law also predicts the bending of light by gravity, which is not surprising given Einstein's demonstration of the "equivalence" of mass and energy. But Newton's equation predicted less bending than Einstein's final more complicated equations. The repeat of Eddington's study that was planned for 2019 might determine whether the "bending" better matches Newton's prediction or Einstein's prediction of approximately twice the "bending". But the most this can prove if they find the larger deflection is that Einstein's complicated equations are more accurate, not that gravity curves space.

In any case, I don't mean in any way to diminish the amazing accomplishments of Einstein. But I do suggest that relativity and quantum mechanics have been interpreted and implemented to reproduce an "observer's world" (as I have already demonstrated above for the Special Theory and for the Copenhagen Interpretation of quantum mechanics), not necessarily the "real world". They provide a mechanics that works for everyday practical calculations for "observers", but not to project properly back through space and time to reproduce the nature, origin and evolution of the universe. This will require a mechanics that recognizes absolute time and space in order to reproduce the "real" world.

The concepts of time dilation and length contraction, as I showed earlier in this chapter, are not real phenomena but errors in observation. I believe a correction and clarification of these flaws in interpretation may help provide a solution to the quantum enigma and allow the further leap forward that Einstein wished to achieve. But before attempting this, I will discuss in Chapter 3 what I consider to be flaws in several of the other "modern" theories of physics, derived based on the Theories of Relativity and quantum mechanics, as well as in the reported experimental and theoretical validations of relativity and quantum mechanics. In Chapter 4, I will summarize the findings and conclusions I have presented in the first three chapters.

CHAPTER 3

CRITICAL ANALYSIS OF THE EXPERIMENTAL EVIDENCE FOR RELATIVITY AND QUANTUM MECHANICS AND RELATED THEORIES

3A STRING THEORY, MULTIPLE DIMENSIONS AND PARALLEL WORLDS

In *The Fabric of the Cosmos* (4, p. 18) Brian Greene claims: *"... superstring theory's proposed fusion of general relativity and quantum mechanics is mathematically sensible only if we subject our conception of spacetime to yet another upheaval. Instead of the three spatial dimensions and the one time dimension of common experience, superstring theory requires nine spatial dimensions and one time dimension. And, in a more robust incarnation of superstring theory known as M-theory, unification requires ten space dimensions and one time dimension – a cosmic substrate composed of a total eleven spacetime dimensions. As we don't see these extra dimensions, superstring theory is telling us that we've so far glimpsed but a meager slice of reality."*

On the other hand, possibly it's proposing a big slice of non-reality! Greene does indeed proceed to rationally suggest *"Of course, the lack of observational evidence for extra dimensions might also mean they don't exist and that superstring theory is wrong."* This would be my take on the situation. But Greene then proceeds to mention that *"String theorists have substantially refined these ideas and have found that extra dimensions might be so tightly crumpled that they are too small for us or any of our existing equipment to see."* The way some put it is that some of these extra dimensions are *"curled up in tiny little balls"* and I can't imagine how these can function as dimensions. The way they are described it sounds like they are tiny little objects that occur in and can be defined by our three real space dimensions and are too small to have any effect on anything except mathematical equations. In any case, some say String theorists are opening up "creative new pathways to reality"; I agree about "creative new pathways", but I am not sure about "to reality". I don't see where they have provided any useful insight into what could constitute a dimension beyond the four dimensions we know and inhabit.

I can only imagine one possibility for an extra dimension that meets my crude definition that it must be a completely independent entity that provides meaningful information about the universe, completely separate from the information provided by the four known dimensions. This could be provided by consciousness, which also provided the observational capability required by quantum theory. An independent view of the universe and the things in it is provided by the brain in the form of its internal pictures and thoughts, which don't seem to exist in the same three space dimensions as the external universe. And the brain's memories provide a view of the past, not provided by the "unidirectional" time dimension of the external universe. Of course, the present state of such memories is very crude, but one can imagine a future when we have evolved into superhuman bio-machines (if the foibles of extreme far right anti-science conservatism doesn't destroy us and our fragile planet first) and such memories become complete and perfect.

Could this constitute a new dimension? I don't know yet – this is the best I can think of as far as a new dimension that isn't a seemingly useless, tiny, curled-up hidden ball whose mathematical representation might be put into an equation to get some mystical prediction. Consciousness could certainly provide meaningful information about the universe, at least its past, and this new dimension might eventually be advanced enough, if we manage to revive the unfairly maligned concepts of reality and determinism, to project the future. Interestingly, this would not be prohibited by the formidable logical argument against the possibility of time travel – we could see back in time to the past, but we couldn't change it. And, furthermore, if we could "see" into the future we might be able to take action that could change it, and this would not violate the logical argument against time travel, either.

On p. 148 of *Parallel Worlds* (6), Michio Kaku says, *"Once we introduce the possibility of applying the quantum principle to the universe, we are forced to consider parallel universes."* We are not **forced** to consider any such thing. Just as there is no evidence for multiple dimensions, there is no evidence for multiple or parallel universes. Furthermore, it probably doesn't make sense to waste time and money ruminating on things that we can never see or validate, as is the case for most of the multiple or parallel world scenarios (although I will present a somewhat different look at this concept in the last two chapters).

These questionable theories all emanate from interpretations of mathematical calculations that don't necessarily have any connection to reality. They are extremely unlikely and beyond our ability to verify or falsify. I will suggest further on, however, a rational scenario for multiple universes (or more correctly, multiple <u>miniverses</u>) that may be testable/falsifiable.

I suggested earlier that the infinite number of solutions so common to theories emanating from quantum mechanics may derive from the infinite number of coordinate systems that are inherent to the Special Theory of Relativity and quantum mechanics. Just as the quantum wave equation generates results suggesting to the converted that particles inhabit all possible positions simultaneously until observation "collapses the waveform" (but common sense informs the rest of us that those positions aren't all occupied simultaneously, but are just the list of all possible positions that the equations were designed to generate), so, applied to the universe, it generates all the possible universes that might be, but only the one we see and inhabit actually exists. In a cycling universe model, some of the more probable of these possibilities could eventually occur as the cycling progresses, but, while we still don't fully understand the creation and meaning of the one universe we can observe, why complicate matters with multiple universes that we can't observe? After all, these multiple universes may be just a product of shortcomings in the interpretation of the mathematical manipulations of the Relativity and quantum theories.

There are some among the quantum enthusiasts who propose an <u>infinite</u> number of "parallel" universes, but this would imply or require an infinite amount of energy and matter, which I understand is impossible. There can be an infinite amount of space and an infinite amount of time (and I believe there must be), but there can't be an infinite amount of matter or energy. An infinite amount of energy in the universe would render the law of the conservation of energy meaningless and unnecessary. Zeno's Paradox allows us to imagine an energy-filled infinite universe, without violating the conservation of energy or providing the impossible (an infinite amount of energy in the universe), but this rescue does not work for an infinite number of parallel universes, assuming each must contain at least some energy (infinity x some energy = infinite energy).

Here are some thoughts from the above discussion: 1. The concept of an infinite number of parallel universes is highly unlikely (in fact, probably impossible) and beyond our ability to confirm even if they did exist (but see Chapters 5 and 6). 2. There is no hard evidence for multiple dimensions beyond the four we acknowledge, other than (possibly) conscious memory, and their existence is highly unlikely. 3. I and others have raised serious questions regarding String Theory and multiple dimensions and multiple (or parallel) universes. 4. Let's continue to try to find a modified or unified theory, other than those mentioned above, to help correct the problems and incompatibilities plaguing relativity and quantum mechanics. I hope that the arguments I provide in this treatise will provide some new directions to take towards this laudable goal, or at least provide an incentive for others more capable than I to pursue this goal.

3B THE BIG BANG THEORY AND INFLATION

To quote Brian Greene in *The Fabric of the Cosmos* (4, p. 14) again: *"It's successes notwithstanding, the theory [The Big Bang Theory] suffered significant shortcomings. It had trouble explaining why space has the overall shape revealed by detailed astronomical observations, and it offered no explanation for why the temperature of the microwave radiation, intently studied ever since its discovery, appears thoroughly uniform across the sky. Moreover, what is of primary concern to the story we are telling, the big bang theory provided no compelling reason why the universe might have been highly ordered near the very beginning, as required by the explanation for time's arrow. …. Inflationary cosmology modifies the big bang theory by inserting an extremely brief burst of astoundingly rapid expansion during the universe's earliest moments (in this approach, the size of the universe increased by a factor larger than a million trillion trillion in less than a millionth of a trillionth of a trillionth of a second)."*

And I can say that I don't believe this absurd suggestion for a millionth of a trillionth of a trillionth of a second!!!

Greene goes on to say: *"… for two decades inflationary cosmology has been harboring its own embarrassing secret. Like the standard big bang theory that it modified, inflationary cosmology rests on the equations Einstein discovered with his general theory of relativity. … physicists have long known that an accurate theoretical analysis of small objects – such as the observable universe when it was a mere fraction of a second old – requires the use of quantum mechanics. The problem, though, is that when the equations of general relativity commingle with those of quantum mechanics, the result is disastrous. The equations break down entirely, and this prevents us from determining how the universe was born and whether at its birth it realized the conditions necessary to explain time's arrow."*

From my perspective, it is impossible to imagine the entire known universe compressed into a single point smaller than a proton (in fact into a volume of zero size) – give me a break! How did that "point" come to be? (I suspect that it couldn't and didn't.) How can it contain the whole universe? How can infinite energy exist in zero volume? (Obviously it can't.) What came before the so-called "Big Bang"? What triggered the bang and fuelled it? And, furthermore, if this impossible thing did happen, there would be a centre to the universe, which apparently there is not.

The Big Bang Theory has serious problems. Inflationary Cosmology was introduced to overcome these problems, but it has even bigger problems and seems very improbable to me. Both theories solve some problems but, in my opinion, introduce even bigger ones. We need some better theories. I elucidated in previous chapters some of the problems I perceive with the prevailing interpretations of the equations of relativity and quantum mechanics and suggested new interpretations. Based on these new interpretations, in the final two chapters I will attempt to provide a start at developing a revised theory regarding the universe and its origins.

3C   THE EXPANDING UNIVERSE

Cosmologists going back to Edwin Hubble have been concluding that the universe is expanding. And by this they don't mean that galaxies are simply moving apart – they mean that space is expanding and pulling the galaxies apart as it does so. And quantum enthusiasts are apparently quite in agreement with this.

But let's think about this for a moment – they are saying that space is expanding! What! First we are told that space is not real and doesn't really exist; we hear *"we entirely shun the vague word 'space' of which we must honestly acknowledge we cannot form the slightest conception"* (9, pp. 10-11), *"Absolute space does not exist"* (4, p. 59). Then we are told that this nonexistent space curves to provide gravity – and, as if that were not enough, we are next told that this nonexistent but curved space is expanding and carrying the galaxies along with it! The cosmologists' observations regarding the so-called expansion of the universe I am sure are correct, just like the equations of relativity and quantum mechanics are correct as far as they go, but I believe there is clearly a problem with the interpretations of what these theories and observations are telling us. I'm not completely alone in questioning the present theory regarding the accelerating expansion of the universe – it has been challenged by others (35, 36).

So, let's analyze this further. I have argued that space not only exists but must be infinite and eternal. And, while space, contrary to the claims of quantum theorists, provides the only absolute benchmark for motion, it has no substance that can expand and exert the force necessary to carry along galaxies. Furthermore, the expansion of something that is already infinite makes no sense at all and would not be measurable! So, this weird conglomerate of contemporary physical consensus that 1) space doesn't exist, but 2) space is expanding and carrying the galaxies along with it (to explain the so-called "expanding universe") makes no sense to me, and I suggest must be wrong. What then is really happening? Here is my take on this:

First, here's the data, the observations. Doppler red shift data suggests that, for the most part, galaxies are moving apart, and the farther away they are from our galaxy the faster they are moving, and this phenomenon is apparently accelerating. To help explain this, the experts use the 2D simplification method and imagine a balloon surface expanding and separating objects on its surface. A reasonable enough suggestion, unless you believe that space does not exist, or, more logically, know that space is not a rubber membrane, but an infinity of nothing and cannot expand, or carry anything along with it even if it were possible for it to expand. Note that a greater red shift for galaxies more distant (i.e., farther back in time because the light has taken longer to reach us) could suggest that the "expansion" is decelerating, not accelerating (i.e., the farther back in time the faster they were separating). And, furthermore, if space is really expanding, and carrying the galaxies along with it, how can the Andromeda Galaxy be rushing towards a collision with our Milky Way

Galaxy, as it is reported to be? Nevertheless, studies of distance-redshift relations of Type 1a Supernovae and other impressive observations are compatible with an accelerating expansion. Therefore, I discuss the so-called "expanding universe" in more detail, and provide a possible explanation for this puzzling concept, including the apparent acceleration, in Chapter 5.

3D   DECOHERENCE

The quantum measurement problem (apparently a consequence of the Copenhagen Interpretation of quantum mechanics) was originally focused on attempting to explain how measuring instruments, which are usually macroscopic objects treatable with classical physics, can give information about the microscopic world of atoms and subatomic particles. Some define the measurement problem slightly differently as the failure to observe macroscopic superpositions, and it has also been phrased that the measurement problem in quantum mechanics is the problem of how (or whether) wavefunction collapse really occurs. These are all somewhat related, but also somewhat disturbing, issues. I suggested earlier that wavefunction collapse is not likely a part of reality. Superposition has been described thus: *"because quantum mechanics is weird, instead of thinking about a particle being in one state, or changing between a variety of states, particles are thought of as existing in all the possible states at the same time"* (see "What is Superposition" at physics.org c/o The Institute of Physics, London), but I discussed earlier the unlikelihood of this being a part of reality. I suggest that the real solution to the "quantum measurement problem" is that, while Planck's quanta are real (and Schrödinger's wave equation works well in most situations, as long as the right coordinate system is used), the quantum mysticism emanating from the Copenhagen Interpretation is not.

Two wave sources are considered coherent if they have the same frequency and a constant phase difference. This is a property of waves that enables stationary interference. It is considered essential for superposition, which is essential for some of the absurdities of quantum physics and will be required for the successful development of the proposed quantum computing revolution. For the qubytes to function as predicted, they will have to be kept entangled in "superposition" and remain coherent. What is considered to differentiate a quantum system from a classical system is the existence of a superposition of states. It has been proposed that when a quantum system is not perfectly isolated, but in contact with its surroundings (the environment), coherence decays with time, a process called quantum decoherence. It is further proposed that decoherence puts a quantum system into an apparently classical state. Decoherence is therefore proposed to form a bridge between the quantum physics of the small and the classical physics of the not-so-small by suppressing quantum interference. But, seriously, if one is going to try to swallow this, one must ask how any quantum system in nature, assuming such could ever exist, could possibly manage to remain "quantum".

As Brian Greene puts it in *The Fabric of the Cosmos* (4, pp. 210-211): *"although photons and air molecules are too small to have any significant effect on the motion of a big object like a book or a cat, they are able to do something else. They continually nudge the big object's wave function, or, in physics speak, they disturb its coherence: they blur its orderly sequence of crest followed by trough followed by crest. This is critical, because a wave function's orderliness is central to generating interference effects.... In turn, once quantum interference is no longer possible, the probabilities inherent to quantum mechanics are, for all practical purposes, just like the probabilities inherent to coin tosses and roulette wheels. ... This suggests a resolution of the quantum measurement puzzle, one that, if realized, would be just about the best thing we could hope for."* He goes on to say: *"Decoherence forces much of the weirdness of quantum physics to leak from large objects since, bit by bit, the quantum weirdness is carried away by the innumerable impinging particles from the environment."* Note that if this were really the case, classical objects would be in a series or continuum of different degrees of coherence between classical and quantum (probably some sort of Bell curve), and this is obviously not the case.

While I find it very hard to stomach these mystic suggestions, Greene continues: *"It's hard to imagine a more satisfying solution to the quantum measurement problem."* I have scoured every classical object in my possession for evidence of a leaking trail of "weirdness" without any success (maybe it would help me if someone could tell me what the basic physical unit of weirdness is).

While I find this proposal very creative, it's hard for me to imagine more decoherent or incoherent weirdness. Greene does go on to say: *"Decoherence allows quantum probabilities to be interpreted much like classical ones but does not provide any finer details that select one of the many possible outcomes to actually happen."* Slightly more satisfyingly, he says further: *"I strongly suspect that there is much insight to be gained by pushing onward toward a complete solution to the measurement problem."* This last statement seems to make a little more sense, since it might allow for a more reasonable solution, such as the one I offered in the last sentence of the first paragraph of this section on decoherence, i.e., that the Copenhagen Interpretation can't be correct and should be reconsidered.

To me, decoherence seems to be best described as wishful thinking pulled out of thin air with no logical understanding or mechanism associated with it. Furthermore, I have argued that the wave equation itself can't interfere, decohere or collapse – it is an equation on paper that predicts the probabilities of what might be observed, and probably nothing more. Going back to the quantum computing dilemma, one must figure out how to keep a "quantum" object "quantum". But, according to the accepted standard Copenhagen Interpretation, such an object doesn't exist until it is observed – and once a "quantum object" is observed it is no longer "quantum"! There is obviously a problem here.

So, again, let's throw away the Copenhagen Interpretation, which I believe doesn't make much sense anyway. Consider also the quantum concept that this "quantum object" exists in all possible positions and states before it is observed – therefore we don't know exactly what it is or where it is, but it is supposedly in superposition and is coherent (of course the quantum concept that an object does not exist at all until it is observed is even more problematic). If we accept this, then it is in the state required for a qubyte, but how do we find it and use it without destroying it? I presume that the brilliant guys developing this technology have already solved or eliminated these issues and I am just behind the times. I will be very impressed if they succeed.

Greene (4, p. 213) mentions that some physicists, of which he is, admirably, not one, claim that: *"To seek an explanation of what's really going on, to strive for an understanding of how a particular outcome came to be, to hunt for a level of reality beyond detector readings and computer printouts betrays an unreasonable intellectual greediness."* I find this statement very disturbing and a terrible attitude to take towards science. My purpose in preparing this treatise is to attempt to reveal and clarify what appear to be quantum absurdities and search for solutions or answers, in plain language rather than esoteric equations, to what is really going on. In other words, to determine the relationship of quantum mechanics to reality and to seek an understanding of the absurdities so quantum mechanics can be properly interpreted and progress towards the more complete theory of nature Einstein longed for. If that is a greedy objective, then I plead guilty. And, if the end result is to prove me, and others who also question the more absurd aspects of quantum theory, wrong, so be it – as long as the proof is based on logical, demonstrable and understandable fact and not merely argumentative opinion.

3E GRAVITATIONAL LENSING AS A TEST OF GENERAL RELATIVITY

It turns out that, in fact, Newton was the first to ask whether the gravitational force acted on light. This was reasonable since he believed light to consist of tiny particles, which therefore probably had a very tiny mass. The German physicist Johann von Soldner in 1804 (38), using Newton's law of universal gravitation, calculated the amount by which light should be deflected if the path of the light ray just grazed the surface of the sun. The deflection angle at the edge of the sun, denoted as alpha, was calculated to be 0.84 arc seconds. Einstein, after developing his Special Theory of Relativity, also calculated the effect that the sun would have on a light ray. Though Soldner's work was apparently unknown to him at this time, Einstein's initial calculation, before he completed the General Theory of Relativity, agreed with Soldner's result based on classical mechanics. After further developing his General Theory of Relativity, Einstein recalculated the deflection angle alpha for light around the sun and found it to be 1.75 arc seconds, approximately twice his and Soldner's previously calculated results. This difference was assumed to be due to the so-called "space-time curvature" effects.

An opportunity to test Einstein's new theory was provided by the total eclipse of the sun in 1919. Sir Arthur Eddington, under the sponsorship of Frank Dyson, the Astronomer Royal, undertook this project. Eddington and Dyson were apparently both supporters of the new General Theory of Relativity. The expedition was beset by many difficulties, but Eddington finally managed to collect some data. Eddington reported confirmation of Einstein's General Theory, but there remains much controversy over these results. Even the larger deflection (1.75 arc seconds) is extremely small, the sun is surrounded by a plasma that will affect the velocity of electromagnetic radiation, the measurements were subject to very large systematic errors and Eddington selected the data he used for the calculations. Marmet and Couture (16) claimed that the results were *"highly unreliable and proving nothing"*. And many other scientists in recent years have questioned Eddington's margin of error, suggesting that his equipment was not sufficiently accurate to discriminate between the predicted effects of the two gravitational theories.

As of this writing, plans are being made to repeat these studies with more modern technology during the eclipse of 2019. It is important to repeat here that Newton's theory and equation also predicted gravitational lensing, i.e., a bending of light by the sun. Einstein's General Theory of Relativity, or at least its equations, only predicted a slightly larger deflection. I don't know which prediction the new studies, if successfully completed, will best agree with, but I believe these results will be irrelevant to the confirmation of the interpretation of the General Theory of Relativity. The verification of the larger deflection predicted by the General Theory of Relativity will not confirm curvature of space by gravity, or should I say the mechanism of gravity as being the curvature of space. The results could only confirm that Einstein and Grossman's complicated equations of General Relativity are slightly more accurate than Newton's original equation, probably because of the inclusion of a few extra factors, such as the mass equivalent of the energy of the gravitational field, as I discussed in the previous chapter. In conclusion, even if deemed apparently "successful", these results will not confirm the interpretation of gravity as being due to the curvature of space.

3F   OBSERVATION OF GRAVITATIONAL WAVES

The first observation of "gravitational waves" was made by LIGO (Laser Interferometer Gravitational-Wave Observatory) on Sept. 14, 2015 (see, for example, "First Observation of Gravitational Waves", Wikipedia) and the second on Dec. 25, 2015 (announced at the 228th meeting of the American Astronomical Society in San Diego). These announcements were greeted with great acclaim as further confirmation of the General Theory of Relativity and validation of the predictions of space-time distortion in the context of large-scale cosmic events.

This accomplishment was described as the observation of the waves given off by the cataclysmic merger of two black holes (thirty-six and twenty-nine times the mass of the sun and colliding at up to 60% of the speed of light) somewhere around 1.5 billion years ago. These waves apparently reached earth as a "ripple in space-time" that changed the length of a 4-km LIGO arm by "a thousandth of the width of a proton", proportionally equivalent to changing the distance to the nearest star outside the solar system by one hair's width. This is clearly an astounding technical accomplishment – hard to imagine possible! The signal was described as in the audible range, lasting over 0.2 seconds, and resembling the "chirp of a bird". LIGO team member Szabolcs Marka, a physicist at Columbia University, claimed *"It's one of the most complex systems ever built by mankind. There are so many knobs to turn, so many things to align, to achieve that [sensitivity]."* Apparently, the experiment is so delicate that unrelated events such as an airplane flying overhead, wind buffeting the building or tiny seismic shifts in the ground beneath the detector can disturb the lasers in ways that mimic gravitational signals. Imre Bartos, another member of the LIGO team at Columbia, claimed, *"If I clap in the control room, you will see a blip."* The mechanism used to attempt to weed out contamination is to have detectors at two separate sites (Hanford, Washington, and Livingston, Louisiana, in this case), but what if the "chirp" is the result of one of the occasional, normally undetectable, underground seismic events occurring somewhere between Hanford and Livingston?

I don't have adequate information and am not qualified to judge the LIGO results, which I consider an impressive technological feat. Nevertheless, considering my previous arguments concerning the impossibility of the "warping" of space as a mechanism of gravity, I would have to suggest that if the "chirps" recorded are not artefacts of some sort, then the explanation has to be something other than "ripples in the curvature of space", such as ripples in a gravitational field. And I wonder how a wave, of whatever nature, generated by the collision of two black holes each 30-40 times the mass of the sun and travelling at 60% the speed of light, could travel 1.5 billion years through so-called "expanding space" and remain a fraction of a second in duration (wavelength)?

3G   GRAVITY PROBE B

Gravity Probe B, though a long and costly exercise plagued by problems, has also been considered by some as providing some confirmation for Einstein's General Theory of Relativity. This NASA mission used four spherical gyroscopes and a telescope, housed in a satellite orbiting 400 miles above the earth, to measure two effects apparently predicted by Einstein's theory of gravity. These were: 1) The geodetic effect – the amount by which the earth is assumed by physicists to "warp the local space-time around it", and 2) The frame-dragging effect – the amount by which the rotating earth is presumed by physicists to "drag its local space-time around with it". These were tested by measuring, with acclaimed great precision, the "displacement angles of the spin axes" of the four gyros over the course of a year and comparing these numbers with the numbers calculated with Einstein and Grossman's equations.

However, despite the claims of success, the probes' data was unexpectedly noisy due to solar flares that interrupted the satellite's observations, and to unexpected torques on the gyroscopes that changed their orientation, mimicking relativistic effects. A review panel claimed that the reduction in noise needed to test rigorously for deviation from General Relativity's equations was *"so large that any effect ultimately detected by this experiment will have to overcome considerable (and in our opinion, well justified) skepticism in the scientific community"* (see, for example, 13, 14).

The above concerns mimic the concerns that have been expressed regarding all of the suggested confirmatory tests of the Theories of Relativity. Many of these experiments have been amazing technological feats but, in my opinion, they all merely attempt to test which set of equations best matches the observations, and don't in any way prove the warping or curvature of space. The present tests show a very slight shift in the position or angle of a gyroscope, extracted out of large technical errors, and which can be mimicked and/or explained differently. For example, earth's rotating mass might cause a slight aberration in a gravitomagnetic field in space (much more likely in my opinion than an aberration in the "curvature of space"), because the field would have energy, and therefore a small mass equivalent. That may be what is being measured and calculated, as provided by Newton's classical law of gravity and more accurately calculated by Einstein's equations.

In summary, the equations of the Theories of Relativity and of Quantum mechanics have been shown to accurately predict what will be observed, but the present interpretations of what the theories are saying seem to lead to absurdities and remain open to questions. These are questions I am trying to answer with the argument that the theories and their equations are in fact calculating what will be observed, but not necessarily what is reality, and that may contribute to the apparent absurdities.

3H QUANTUM ENTANGLEMENT AND BELL'S THEOREM AND INEQUALITY

According to Wikipedia: *"**Quantum entanglement** is a physical phenomenon that occurs when pairs or groups of particles are generated or interact in ways such that the quantum state of each particle cannot be described independently of the others, even when the particles are separated by a large distance – instead, a quantum state must be described for the system as* *a whole."* Further on this article states, *" ... any measurement of a property of a particle can be seen as acting on that particle (e.g., by collapsing a number of superposed states) and will change the original quantum property by some unknown amount; and in the case of entangled particles, such a measurement will be on the entangled system as a whole. It thus appears that one particle of an entangled pair* **'knows'** *what measurement has been performed on the other, and with what outcome, even though there is no known means for such information to be communicated between the particles, which at the time of measurement may be separated by arbitrarily large distances."* This kind of statement, referring to particles as "knowing" things, as if particles had brains, is common in articles dealing with quantum mechanics.

Brian Greene in *The Fabric of the Cosmos* (4, p. 112) claims: *"Through the Heisenberg uncertainty principle, quantum mechanics claims that there are features of the world – like the position and the velocity of a particle, or the spin of a particle about various axes – that cannot simultaneously have definite values. A particle, according to quantum theory, cannot have a definite position and a definite velocity; a particle cannot have a definite spin (clockwise or counterclockwise) about more than one axis; a particle cannot simultaneously have definite attributes for things that lie on opposite sides of the uncertainty divide. Instead, particles hover in quantum limbo, and a fuzzy, amorphous, probabilistic mixture of all possibilities; only when measured is one definite outcome selected from the many. Clearly, this is a drastically different picture of reality than that painted by classical physics".* **Yes, thank goodness for classical reality – most of the above makes no sense, as I have repeatedly shown.**

Brian Greene continues: *"Ever the sceptic about quantum mechanics, Einstein, together with his colleagues Podolsky and Rosen, tried to use this aspect of quantum mechanics as a weapon* [**rather than "weapon" I would say "logical argument"**] *against the theory itself. EPR (Einstein, Podolsky, Rosen) argued that even though quantum mechanics does not allow such features to be simultaneously determined, particles nevertheless do have definite values for position and velocity; particles do have definite spin values about all axes; particles do have definite values for all things forbidden by quantum uncertainty. EPR thus argued that quantum mechanics cannot handle all elements of physical reality – it cannot handle the position and velocity of a particle; it cannot handle the spin of a particle about more than one axis – and hence it is an incomplete theory."* Yes, and I believe EPR was right and have tried to show throughout this book that the quantum world is the observed world, not the complete real world. And Einstein was certainly right that it is taking Heisenberg's Uncertainty Principle too far to claim that a particle can't have a definite position and a definite velocity as opposed to the actual principle that we can't measure both with precision greater than the uncertainty stipulation.

EPR (42) was actually a thought experiment that exposed what Einstein called the "spooky action at a distance" accepted by the standard interpretation of quantum mechanics. EPR embodied a serious challenge to this interpretation and upset Bohr who struggled to oppose it but ended up publishing a rambling incoherent counter-attack (43) that left things just as muddled. Bohm and Aharanov (44) later presented a variation of the EPR experiment. Paul Halpern in *Einstein's Dice and Schrödinger's Cat* (12, p. 209), an excellent book by the way that I enjoyed very much, gave a succinct summary of this experiment that I will try to dissect. Halpern states: "*The Bohm-Aharonov version of the EPR thought experiment imagines 2 electrons from the same energy level propelled in different directions.* **Pauli's exclusion principle guarantees that the electrons must have opposite spin states: if one spin is up, the other spin is down**. *Until a measurement is taken, it is impossible to know which is which. Therefore* [according to quantum theory], *the 2 electrons form an entangled quantum state that is an equal mixture of both possibilities: up-down and down-up.* [No - the electrons' spins are fixed (one 'up' and one 'down'– once we measure one we know the other – there is no quantum mixture except in the **observers' minds**.] *Now suppose an experimenter measures the spin of one of the electrons using a magnetic apparatus and another researcher immediately records the spin of the other. According to the orthodox quantum interpretation, the* system *would instantly collapse into one of its spin eigenstates, either up-down or down-up.* [**No - there is no collapse and no mixture of eigenstates – you measure one spin and you know the other, period!**]. *So, if the first electron's reading was spin up, the other would automatically be spin down* [**YES! Brilliant!**] *In the absence of an interaction through space between the two, how would the second electron instantly "know" what to be?"* In fact, the electron didn't "know" what to be, it just was - it was pre-ordained and obvious without any quantum spookiness, regardless of Bell's Theorem and Inequality (45), which I will now discuss briefly.

It was the eventual testing (e.g., 46) of Bell's Theorem and Inequality that tipped the scale back towards the Copenhagen Interpretation. The existence of "hidden variables" was proposed as a solution to the puzzle of the apparent "action at a distance" that Einstein called "spooky". Bell's Theorem was set up to test whether "hidden variables" could account for the apparent non-locality. In Bell's defence, he was attempting to show that quantum mechanics was clearly inadequate, as was believed by Einstein and Schrodinger, and he was surprised, as were those who conducted the subsequent tests, when it seemed to show the opposite. What Bell's Inequality seemed to suggest, for those who fully accepted the experimental data (regardless of loopholes, etc), was that reality and locality (an object is directly influenced only by its immediate surroundings) cannot both exist. It didn't say that neither can exist. And, as I have shown throughout this treatise, in agreement with Einstein, there is and must be reality. If there was no reality there would be nothing to observe. Indeed, I am confident that the description of reality as inherently nonlocal and created by observation is not correct and quantum mechanics (while correct as far as it goes) is incomplete as Einstein, Schrodinger and many others have contended. Therefore, I believe there is a flaw in the implications of or interpretation of Bell's Theorem and Inequality. I won't attempt to present the whole mathematical exercise involved in demonstrating this here as it is complicated and esoteric, and has already been done by Dr. Joy Christian (47) and others.

Dr. Christian's work refuting Bell's Theorem has been itself refuted but he has responded quite adequately to these attacks. Following is a portion of a very succinct review of Dr.

Christian's published book (titled "Disproof of Bell's Theorem: Illuminating the Illusion of Entanglement"): "… in this ground-breaking collection of papers, the author exposes a fatal flaw in the logic and mathematics of Bell's Theorem, thus undermining its main conclusion, and proves that --- as suspected by Einstein all along --- there are no spooky actions at a distance in nature. The observed long-distance correlations among subatomic particles are dictated by a garden-variety "common cause", encoded within the topological structure of our ordinary physical space itself." I tend to agree.

This is still a topic of controversy that needs to be settled, but let's be clear – the standard interpretation of quantum mechanics yields some absurdities that defy common sense. The believers say that if the data suggests that these absurdities are right, then we must accept it as trumping common sense. Well, all the data I have seen generated along this vein comes from amazing, but very complicated, technical accomplishments that yield results and conclusions that I believe are open to question or reinterpretation. On the other hand, the data generated by common sense observation and logic is very uncomplicated and compelling. My favourite is Einstein's *"I can't believe the moon is not there when I am not looking"*. This example was more for comic shock value, but I have mentioned many more detailed examples throughout this treatise. I will repeat one more time – relativity and quantum mechanics are very successful as far as they go because they calculate what will be observed from a defined reference frame, but what has not yet been observed will still exist and have position, velocity, etc. in reality. Therefore, quantum mechanics is not complete because it cannot recognize the existence of the unobserved reality.

CHAPTER 4
SUMMARY OF FINDINGS AND CONCLUSIONS

I have presented arguments throughout this treatise for the following concepts:

It seems obvious that space exists, space is real, and the only real independently existing trajectory is the trajectory in space. Since there are an infinite number of possible surrogate rigid bodies of reference from which to make observations, there are an infinite number of such surrogate trajectories. Einstein's Special Theory of Relativity allows a correct determination of what will be observed from each such trajectory, but not an extrapolation back to a "beginning" because that requires the extrapolation back of the real independently existing trajectory in space and time, and the theory doesn't recognize such a trajectory. And, furthermore, there was no beginning of time or space, only of matter and energy.

Regarding his classic "embankment and moving train" example, Einstein claimed: *"Events which are simultaneous with reference to the embankment are not simultaneous with respect to the train, and vice versa (relativity of simultaneity). Every reference body (co-ordinate system) has its own particular time; unless we are told the reference-body to which the statement of time refers, there is no meaning in a statement of the time of an event."* I, on the other hand, make the claim that **there is time and there is the perception of time** (relativity pertains to the perception/observation of time, not to time itself), and suggest the following correction to his statements: "Events which are <u>observed</u> to be simultaneous with reference to the embankment do not <u>appear</u> to be simultaneous when <u>observed</u> from the train, and vice versa (<u>relativity of the **observation** of simultaneity</u>). Time is progressing identically

59

everywhere, but every reference-body (co-ordinate system) provides its own assessment (observation) of the timing of events elsewhere: unless we are told the reference-body to which a statement of observed time refers, we cannot assess its relationship to the time we observe from another reference system" (the relativity of the **observation** of time).

Space and time are real, absolute and independent – but the observations of space and time are relative. Time is (and must be) progressing identically everywhere and every reference frame has (and must have) the same time, but there may be an error in the observation from one reference frame of the timing of events in another reference frame.

It has been claimed that the Theories of Relativity and quantum mechanics are correct new laws of nature, while Newton's laws only work for a limited special case. It has also been claimed that the Theory of Relativity toppled Isaac Newton's physics and redefined our notion of space and time. I have proposed alternatively that the Theories of Relativity and quantum mechanics have not displaced Newton's laws of nature – Newton's laws are still the only true laws of nature. The Theories of Relativity and quantum mechanics have merely rephrased Newton to correct for very special circumstances (for example, for motion approaching a significant fraction of the speed of light in the case of Special Relativity, and in the case of General Relativity, taking into account the fact that gravitational fields act on gravitational fields, as well as a very complicated effect of gravity on an object's **observed** trajectory due to acceleration with its changing velocities).

Contrary to the prevailing belief, I claim that time and space are both infinite and eternal, and indeed must be; there is no other acceptable alternative. Quantum physicists, furthermore, say that time seems to have a direction pointing from past to future for which neither the Theories of Relativity nor quantum mechanics has provided an explanation. Indeed, these theories contain no distinction for direction of time – they work identically in both directions. This is not surprising since the Theories of Relativity and quantum mechanics are based on an infinity of "rigid coordinate systems" that provide no specific direction. This indicates that the Theories of Relativity and quantum mechanics have a serious problem and are not complete. Quantum physicists also ask what conditions at the creation of the universe imprinted such a direction on time. I claim alternatively that time has an obvious direction that has an intuitively obvious reason. Nothing imprinted a direction on time – **time imprinted a direction on reality and the evolution of the universe and the Theories of Relativity and quantum mechanics have not captured this fact and are not complete** (see note at end of this chapter and Appendix 9 for further comments on the quantum physics concept of time).

It is recognized that the Big Bang Theory and Inflation, derived based on the Theories of Relativity and quantum mechanics, have serious problems, to say the least. The thought of the universe beginning as a point in space ("smaller than a fraction of the size of a proton", or, even worse, of zero volume and containing an infinity of energy) and inflating 1 million trillion trillion times in less than 1 millionth of a trillionth of a trillionth of a second is clearly impossible to believe. When these theories are combined to project back to a supposed "beginning", they end in an impossible singularity. It seems, therefore, that the Big Bang and Inflation theories cannot be correct, that is, they never occurred <u>as described</u>. To repeat, in fact nothing imprinted a direction on time – rather time imprinted a direction on reality and the evolution of the universe.

I, furthermore, provided arguments that the Special Theory of Relativity is not a new law of nature, but merely a correction for observation (based on the Lorentz Transformation), required because the speed of light, required for observation, is finite rather than infinite (that is, information transfer is not instantaneous). If the speed of light were infinite there would be no correction required.

Many, including Einstein, concluded that in the Theory of Relativity the velocity of light c plays the part of a limiting velocity, which can be neither reached nor exceeded by any real body. This is based on the observation that in the Lorentz Transformation, when the velocity v equals c, the transformation becomes infinite, and for still greater velocities the transformation becomes imaginary. I argue that this is not the proper conclusion – it may be true based on other considerations than the Lorentz Transformation, but all the Lorentz Transformation says is that observation is not possible when the velocity v reaches or exceeds c.

I have, furthermore, demonstrated that length contraction and time dilation are not real, that is, do not actually occur. They are just erroneously perceived to from a reference frame in motion relative to the object in question (**relativity of observation**). And the so-called experimental proofs of these observational phenomena, don't actually prove what they claim to.

The generally accepted interpretation of the General Theory of Relativity is that gravity curves/warps space and matter follows this curve. I have argued that this is complete nonsense. Space is nothing and has no substance that can be curved. No mechanism has been suggested for such an action and even if space could be curved, that wouldn't explain gravity. The explanation for something falling down a curve is gravity – so the effect of gravity is explained by the effect of a curve, which is explained by the effect of gravity. This is, of course, circular reasoning.

I suggest that the curvature the equations describe is not a curvature of space, but simply the curved trajectory of the object in space. The ten complex nonlinear equations were generated (by Einstein and Grossman) to calculate the correct trajectory for an accelerated object. This probably begins with the Lorentz Transformation correction for a body moving at velocity v, but in the case of an accelerated object (for example an object in a gravitational field), the velocity is constantly changing, which requires very complex corrections. The complex equations required must also compensate for the iterative production of gravitational fields due to gravitational fields producing gravitational fields. I bolster this argument with the elucidation that the curvature of space cannot possibly explain the formation of white dwarfs or neutron stars, and many other effects of gravity.

The generally accepted Copenhagen Interpretation of quantum theory claims that observation creates reality – before observation, the object is a nebulous probability waveform existing everywhere but nowhere. According to Niels Bohr, until an observation or measurement is made, a microphysical object like an electron does not exist anywhere. I argue that that this is nonsense. In the real world an object exists in an exact position that the observer cannot know until he/she makes an observation. The quantum equation (describing the waveform or wave function) describes the probability of all possible positions for the object. But once the observer makes an observation, all the possibilities in the equation (and in the observer's mind), except the one he/she observes, become meaningless. The observer has placed a piece of information not in reality but in his/her observer's world (i.e., the quantum world) – nothing collapses or is created in the real world.

I have shown that the data generated by the various tests supposedly confirming the Theories of Relativity, such as measuring gravitational lensing, gravitational waves and the Gravity Probe B, among others is not overwhelmingly convincing (although these were amazing technological achievements). I have further argued that even if one accepts the data as satisfactory, what this data shows is merely that Einstein's more complex equations are more accurate than the original simpler equation of Newton, not that the standard interpretations of what these theories mean are correct. I have, furthermore, explained why this is and have provided different interpretations for these theories. Many authors have touted that Einstein's Theories of Relativity predict the bending of light by gravity and that this has been confirmed. Fine - but they often neglect to mention that Newton's theory also predicts the bending of light by sufficient gravity – there is only a quantitative difference in the amount of bending predicted. Einstein's equations may be more accurate, for the reasons I stated above, but there is no proof for curvature of space in any of the studies reported.

It should be completely obvious, even to a non-physicist, that the laws of nature must be the laws of nature regardless of the coordinate system in play or the state of motion of an object or a frame of reference. It also seems obvious that, since observation requires light waves, or other electromagnetic waves, and the speed of these waves is a finite constant, not infinite (that is, information transfer is not instantaneous), that there will be observational delays between observers at different distances or different speeds relative to the observed. Therefore, mathematical corrections may be necessary between different frames of reference. This in fact, not surprisingly, turns out to be the case. These corrections do not represent changes in the laws of nature, but merely obvious corrections that can be determined and applied using simple mathematics easily envisaged by and consistent with classical mechanics and its laws of nature. The required correction turns out to be the Lorentz Transformation, and this is not a new law of nature but merely a correction factor for observation between different frames of reference, required due to the finite speed of light.

The Special Theory of Relativity provides a modification of classical mechanics, applying the Lorentz Transformation to correct for the error in observation at speeds at which the ratio v^2/c^2 becomes significantly greater than zero (this being required because the velocity of light c is not infinite). If the velocity of light were infinite, there would be no correction needed; the Lorentz Transformation (correction) would reduce to 1x. The transformation becomes infinite or meaningless if v reaches or exceeds the value of c. I mentioned above that some, including even Einstein himself, have claimed that this means that the velocity of light c cannot be exceeded, but while it may be true that the velocity of light c cannot be exceeded, this must be determined by some other evidence. All the Lorentz Transformation logically means here is that observation is impossible at speeds exceeding the velocity of light. The "**observer's world**" disappears at velocities that equal or exceed that of light, but of course the **real world** remains unchanged.

While this correction/modification to classical mechanics provides practical benefit to certain observations, it also introduces serious problems in defining the "real world" and developing a unified theory that can incorporate time and space correctly in order to calculate back to the "beginning", and thereby understand the origin and evolution of the universe. It also tends to introduce multiple nonexistent dimensions and infinite numbers of solutions, and ends up producing theories that predict things that appear absurd and that can't be experimentally corroborated. However, due to physicists being devotedly wedded to the standard interpretations of relativity and quantum mechanics (somewhat understandable because of the significant areas of success for these amazing creations of the 20th century), they apparently don't spend enough time questioning these absurdities and looking for new directions. Rather, they continue to devote the majority of their man-hours to working on and writing about theories that emanate from these misinterpretations (as mentioned in ref. 2).

Unfortunately, the standard interpretation of quantum mechanics is so disturbingly unlikely that I am completely dumbfounded that so many brilliant folks accept it. There is no way that observation creates reality (other than our perturbed perception of reality). I hope I have presented enough arguments to at least convince some physicists to have second thoughts about the standard interpretation of quantum mechanics and the prevailing interpretations of the Theories of Relativity, or at least to attempt to generate convincing evidence that proves me wrong.

Heisenberg's Principle of Uncertainty certainly has an element of reality to it – there is no question that we have limitations to our technical abilities and our observational acumen, and there is also no doubt that observation can influence the observed. Of course, it has been stated, *"the uncertainty principle actually states a fundamental property of quantum systems* [note: there is uncertainty in the quantum world because the quantum world is the observed world, not the real world, and information transfer is not instantaneous] *and is not a statement about the observational success of current technology"* (25), and uncertainty is suggested to arise in quantum mechanics due to the matter-wave nature of quantum objects (24, 25). Indeed, the double-slit experiment suggesting the dual matter-wave condition is definitely a serious issue that needs better clarification. A proper explanation of this puzzling finding will certainly advance our understanding of nature and help develop a more complete quantum physics. I have discussed the extreme unlikelihood of the reality of the "collapse of wave functions" and of the suggestion that each particle exists everywhere and/or nowhere until observed. Indeed, quantum mechanics describes the "observed world", and there is uncertainty in the "observed world", but not in the "real world". I suspect that Heisenberg's "uncertainty" would disappear if information transfer were instantaneous (i.e., if the speed of light were infinite).

Furthermore, I would have guessed that the more precisely we determined the location of a particle, the more precisely we would be able to determine its velocity (see thought experiment below, Fig. 9) – but that is just logic. Proper experimentation trumps logic, if the experiment is conducted logically. Let's assume Heisenberg met these criteria and we truly can't determine with precision both the location and the velocity of a particle. Looking at the equation Heisenberg proposed to introduce the concept ($D_x \times D_p > h/4pi$), it is obvious that uncertainty is built into this equation. But, in any case, accepting the possibility that we cannot precisely determine both the position and the velocity of a particle (**observation**) in no way indicates that the particle does not have a precise position and velocity (**reality**). Furthermore, there is no way this concept, or any of the other mystical quantum concepts developed for the micro-world, carries over into the macro-world, although many have tried to make this leap (and the fabricated concept of decoherence does not seem to me to save these absurdities). I hope I have made these thoughts clear in the previous chapters.

Figure 9

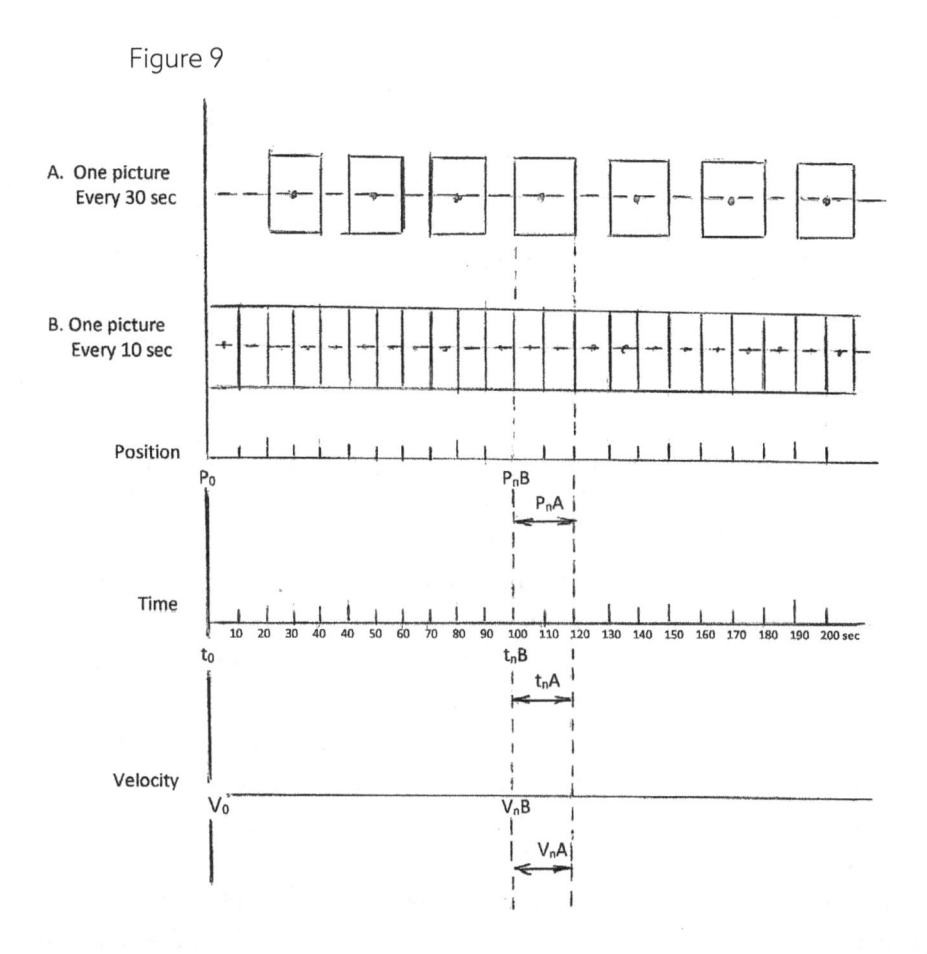

Thought experiment supporting the suggestion that the more accurately we can determine the position of a particle, the more accurately we can determine its velocity. Consider a burst of photos of an accelerating object – the more photos we have (providing finer time definition), the better we can determine position, and the more accurately we can determine the position, the more accurately we can determine the velocity.

Note that Position $P_n = v_0 t_0 + 1/2 a t_n^2$ and Velocity $v_n = v_0 + a t_n$. Note further that v_0, t_0 and a are known and that by determining P_n we can calculate t_n from the first equation above. And, knowing t_n we can calculate v_n. Consider now the following:

Photo sequence A: one photo every 30 seconds, each photo taking 2 seconds to complete. Photo sequence B: one photo every 10 seconds, each photo taking 10 milliseconds to complete.

Clearly we can determine position P_nB more accurately using photo sequence B than position P_nA using photo sequence A. Therefore, we can calculate t_n more accurately, and by using a more accurate t_n we can determine v_n more accurately. <u>Therefore, the more accurately we determine position, the more accurately we can determine velocity.</u>

Now let me continue with the summary of what I have demonstrated in the previous chapters.

A problem, as I mentioned earlier, arises from the denigration of space and time in the Theories of Relativity and quantum mechanics and also from the replacement of the actual benchmark of space with an infinite number of surrogate "rigid" coordinate systems. On the other hand, the equations associated with these theories are able to calculate with accuracy what will be observed from a given defined coordinate system. It is this utility that keeps the standard quantum theory alive and kicking in spite of the absurdities it harbours – but to quote Brian Greene (4, p10). "… *utility and reality are very different standards.*".

Mathematics is a powerful tool, but it is only that, a tool, and can predict nonsense. Time provides a fourth dimension and it does not require the Theory of Relativity to recognize this. Furthermore, time is independent and absolute. It is only the observation or measurement of time that is relative, not time itself. Unfortunately, there is no primary origin for time (i.e., $t = 0$), and this provides a complication. Space exists; space is real and important and provides the benchmark for motion. Unfortunately, there is no primary origin for space (i.e., $x = 0$, $y = 0$, $z = 0$), and this also provides a complication. Nevertheless, there is such a thing as an independently existing object trajectory. Replacing it with a trajectory relative to a specified rigid coordinate system allows a correct calculation with respect to that body of reference, but this concept replaces the single reality with an infinite number of potential observational surrogates, and prevents these theories from being extrapolated back to provide any rational theory of the origin of the universe.

As I have stated previously, the Special Theory of Relativity is not a new law of nature but a mathematical correction necessary to validate an observation from the chosen frame of reference. Throughout relativity and quantum mechanics, the real world is replaced by the observed world. Contrary to Einstein's claims during the development of the Special Theory of Relativity, I have shown that simultaneity occurs in reality, and can be demonstrated and confirmed with simple mathematics consistent with both relativity and classical mechanics. Furthermore, every reference frame does not have its own time, as has been claimed even by Einstein himself (9, p. 31), but merely its own perception of time, All reference frames have the same time, but the perception of time in a different reference frame will be altered because of the finite speed of light, and the Lorentz Transformation provides the required correction that matches what will be observed. No wonder the theories break down in singularities when one tries to trace them back in time, since the theories give a different time to every frame of reference, and don't recognize the unidirectionality of time – nor do they have an origin for the benchmark of space.

The Special Theory of Relativity is a useful correction of the basic classical law, for certain extreme situations where correcting for observational error is required, because the velocity of light is less than infinite. The theory has been so successful because it indeed corrects for observational error, matching observation for a given reference frame. But, contrary to the standard Theories of Relativity and quantum mechanics, space and time are real and are infinite and eternal. They must be infinite and eternal. There is no other possibility. If we try to replace empty space with nothing what do we get? Nothing, of course, which is empty space.

There is no "fabric of spacetime". There is space and there is time. Spacetime is an artificial construct that may curve with gravity (acceleration) because with motion position changes with time. Under acceleration the amount of change per unit time increases or decreases with time, yielding a curve – not a curvature of space, but a curvature of the trajectory of the object being tracked. Neither time nor space can be created or destroyed.

Time is meaningless to empty space. While they have absolutely no effect on each other, space could not exist without time and time could not exist without space. Once matter and energy appear, time becomes important to the motion and evolution of matter, but there is nothing to recognize and experience this until consciousness evolves. There is no meaning to time without consciousness. But consciousness just recognizes the present; it is not until intelligence evolves that the past and the future are recognized and the importance of time to motion and evolution are understood.

We cannot see time. Time has no substance or physical properties that we can measure. We observe things change and that gives us the feeling and a measure of time. There is no real absolute time, because there is no origin, no "beginning". It "arrives" from infinity, or better stated, "eternity". There is no absolute $t = 0$ because time is infinite and eternal. It is extremely important to understand these concepts. Length contraction and time dilation – which are the product of the Lorentz Transformation, the essence of the Special Theory of Relativity – are not real. They are merely corrections for errors in observation with respect to specified coordinate systems, required because the speed of light is not infinite, but finite and constant in all frames.

The original development of the theory and equations of relativity was claimed to be based on two postulates:

1. The laws of physics are the same in all frames of reference (please note that I removed the term inertial and I am trying to make a distinction between laws and mathematical equations).
2. The speed of light in vacuo has the same value in all inertial frames of reference.

But this approach was simply a clever way of forcing derivation of the Lorentz Transformation into appearing to be a new "law" of sorts. The first postulate is obvious, as I mentioned earlier, and the second "postulate" had the utility of requiring the application of the Lorentz Transformation to the mathematical determination of times and lengths between the different frames of reference, and making it seem like a new law had been derived. As I have already stated, Special Relativity simply corrects calculations to what will be observed from a given reference frame. The Lorentz Transformation, the essence of Special Relativity, disappears with the introduction of instantaneous information transfer (i.e., if the speed of light were infinite) and I suspect the same thing would happen with Heisenberg's Uncertainty Principle and Bell's inequality. I suspect, furthermore, that Bell's inequality merely confirms that quantum mechanics calculates what might be observed, not that reality and/or locality do not exist (see ref 46 for a strong disproof of Bell's Theory and Inequality).

I have attempted to show that the General Theory of Relativity is not a revolutionary new law that explains gravity. In fact, it is a revolutionary and complicated set of equations that accurately calculate the trajectory in space of an object experiencing gravity. Gravity does not and cannot curve space, and that wouldn't explain gravity even if it did. Gravity appears to work in a straight line between the centres of the two masses concerned. And "things" follow "curved geodesics" only if the straight-line trajectory is not accessible (such as, for example, airplane and road vehicle routes around the "globe") and this is totally compatible with classical mechanics. And space is apparently "flat", according to all careful observations (see WMAP 2013 results).

I want to emphasize that we obviously don't yet fully understand gravity – the presently accepted interpretation of the General Theory of Relativity is clearly not correct. This is very important to recognize as I believe that a proper understanding of gravity will eliminate the locality/nonlocality issue, explain dark energy, allow a better understanding of the universe and its origin, and may eventually allow galactic and possibly even intergalactic travel.

We cannot discard the concepts of absolute time and space. Contrary to the accepted dogma since the advent and acceptance of the Theories of Relativity and quantum mechanics, time is real, independent and absolute and can't be created or destroyed, but is infinite and eternal. This seems obvious to me. The equations of relativity and quantum mechanics are shown to be incompatible when attempts are made to combine them to peer back to the "beginning" or to explain the very small. They produce singularities and impossible theories like the Big Bang Theory and Inflation when they are used to try to delve into the origin and evolution of the universe.

The reason for this is that time is eternal (an interesting form of infinity) – therefore there is no beginning, no $t = 0$. Time is uniquely unidimensional and unidirectional. We can establish an artificial origin and project forward towards infinity, but, looking backwards (we can look backwards but not go backwards) provides a confusing dilemma. It is easy to imagine infinity going forward but not so going backward. We can't establish where (or should I say when) we are because there was no beginning or origin (due to the fact, that time is eternal and must be so). It is impossible to imagine or enumerate when we are since we are at an infinite time from the "beginning", and indeed, because of this, there was no "beginning". And this explains why physics and cosmology are providing no insight regarding the "beginning". There was no beginning of the universe, only a beginning of matter in the universe. The existence of matter allows us to establish surrogate "rigid coordinate systems" to facilitate calculating what we will observe, but not to discern reality unless we understand that space and time are real and provide the only absolute benchmarks for motion. We must dispense with the dogma and learn how to use these benchmarks for reality if we are to discern the origin and evolution of the universe.

The suggestion that the flow of time can be affected by some little event (and indeed, if this were possible, there would be a chaotically infinite number of such events) in one of an infinite number of surrogate "rigid coordinate systems" is patently ridiculous. The perceived differences between different frames of reference are merely that, perceptions, or shall we say observational errors that can be corrected for by applying relativistic equations (i.e., Lorentz Transformations).

Regarding Einstein's "falling-man thought" and the Principle of Equivalence, I have argued that free fall towards the earth is not the same as the experience of floating in space devoid of all gravitational influence – this is a complete fallacy. This is another example of "observation" versus "reality". The "falling man" may not feel the force on him (observation) because there is nothing pushing back yet, but he is indeed experiencing a force and being accelerated towards the earth (reality) – he is building up kinetic energy that he will become aware of when he reaches earth. This kinetic energy does work, dramatically expressed as mechanical destruction (broken bones and tissue) and eventually ending as thermal energy (an example of an "observer" finally recognizing "reality" – but much too late and unpleasantly).

There is no evidence for multiple dimensions or multiple or parallel worlds, and I have argued there is no sense spending time on things we can never observe or validate (although I provide a possible alternative to this declaration in Chapters 5 and 6). And I also argued that the infinity of solutions so common to theories emanating from relativity and quantum mechanics occur because of the infinite number of potential coordinate systems that are inherent to the Theories of Relativity and quantum mechanics. Regarding the expanding universe, I argued that while I expect the cosmologists' observations are correct (just like the equations of relativity and quantum mechanics are correct as far as they go), there is clearly a problem with the interpretations of these observations. The universe is not expanding, i.e., space is not expanding and can't expand – rather galaxies are separating. And I have suggested that the problem is that these are all theories of the "observed world", not the "real world". I will in the next chapter provide different explanations for the origin and apparent accelerating expansion of the universe.

I discussed in the last chapter the problems with the Big Bang Theory, Inflation and String Theory. It's impossible to accept the concept of the whole universe compressed into a single point immensely smaller than a proton (in fact of possibly zero volume), and this infinitesimal point containing an infinite amount of energy. It is just as impossible to accept the rate of expansion proposed by the Big Bang and Inflation, particularly with no idea of what was before this or how or why this happened or what powered it. Furthermore, if the universe started as a point, there would be a centre to the universe and therefore an origin for time and space. But apparently there isn't. None of these issues plague the theory I will present in the next chapter, and, furthermore, this new theory might provide an explanation for "dark energy".

Note: For more than a century now, I suspect that theoretical physicists have misunderstood and misrepresented space and time. That is the reason Scientific American editors were able to claim correctly that theoretical physicists have not produced a single useful new theory of nature in that whole period of time (1a). What theoretical physicists don't seem to realize is that time and space are real, absolute, infinite and eternal but have no physical substance or properties that they can measure or affect. Their laws and equations and observations have dealt only with the physical properties and relations between matter and energy that, unlike space and time, had a beginning and will likely have an end. Physicists have developed amazing material instruments to measure distance and duration (substitutes for space and time) between material objects and events with incredible precision, but they are not measuring space or time; space and time have no substance or properties that can be measured or influenced.

CHAPTER 5

A STAB AT A NEW THEORY OF THE ORIGIN AND EVOLUTION OF THE UNIVERSE

A useful introduction to what I want to say in this chapter is provided by Martin Bojowald in his introduction to his book *Once Before Time: A Whole Story of the Universe* (33). He states:

"... *questions that have engaged thinkers over the centuries, and that remain of great significance, in quantum theory as well as general relativity, are the role of observers in the world and the question of what can be observed at all and what perhaps cannot. ... ... The big bang model is founded on general relativity – as a description of space, time and the driving gravitational force – as well as on quantum theory ....*

"*In spite of all their successes, general relativity together with quantum theory, as they are being used today, do not provide a complete description of the universe. When one solves the mathematical equations of general relativity in hopes of finding a model for the temporal evolution of the universe and its long term history one always reaches a point – the so-called Big bang singularity – where the temperature of the universe was infinite. ... But infinity as the result of a physical theory simply means that the theory has been stretched beyond its limitations, its equations lose all meaning at such a place. In the case of the big bang model, one should not misunderstand the breakdown of the equations as the prediction of a beginning of the world, even though it is often presented in this way. ... what happened at the infinity of the big bang singularity …? Was this really the origin of the world and of time, or was there something before? And if there was something before the big bang, what was it?"*

Of course, there was something before the Big Bang, assuming it ever happened as suggested. I will attempt to provide an answer, but it may be hard to accept for those so heavily invested in the present impressive, but inadequate, theories. Of course, what I propose may be wrong and I hope those with sufficient knowledge and insight will demonstrate clearly with demonstrable facts where I might be wrong, and also hopefully where I might be right.

To begin the process of suggesting a different explanation/theory of what the universe is and how it came to be, let's start with Newton's laws of motion and try to make something clear:

First law (the law of inertia): Bodies move in a straight line with a uniform speed, or remain stationary, unless a force acts to change their speed or direction.

Second law: A force produces an acceleration (a) that is in proportion to the mass (m) of the body acted upon, and in the direction of the net force (F = ma, where F = net force).

Third law: Every action of a force produces an equal and opposite reaction.

Newton's law of universal gravitation: a particle attracts every other particle in the universe with a force that is directly proportional to the product of their masses and inversely proportional to the square of the distance between their centres.

These simple but brilliant laws still hold almost perfectly to this day. They form the basis of classical mechanics, and even of relativity and quantum mechanics with only some modifications for some special conditions.

While physicists claim that the Theories of Relativity and quantum mechanics are the most successful theories yet devised, based on their successful application of its equations to various problems, I have argued that, in fact, these equations merely apply certain corrections to classical mechanics for errors in observation due to lack of instantaneous information transfer. They reproduce what will be observed, i.e., they reproduce the "observer's world" rather than the "real world". There is only one "real world" but there are an infinite number of possible "observer's worlds", depending on the observer and the coordinate system used for the observation. The Special Theory of Relativity provides a modification to classical mechanics to correct the calculation to what will be observed from a given coordinate system. Unfortunately, while the equations of relativity and quantum mechanics have a very practical utility, providing accurate calculations of what will be observed in many practical present-day situations, in my opinion they mislead us to some degree when we accept the present interpretations of what these equations mean and try to extrapolate them towards an origin for what we consider to be the universe.

So, let's try some thought experiments to explore an alternative analysis of how we came to be. I have made the point in previous chapters that the "revolutionary new ideas regarding space and time that toppled Newton's physics" have misled us and slowed our progress towards understanding how we came to be. In fact, there is space and there is time. Both are absolute, independent, infinite and eternal. It is obvious with a little thought that they must be, there is no other alternative – I argue that they can't not be. They can be neither created nor destroyed. They are the crucible of the stage onto which we walked after billions of years of preparation and evolution. Time is certainly meaningless to empty space, but once matter comes along, time becomes important to the motion and evolution of matter within space. Unfortunately, there is nothing to recognize and experience this until consciousness is evolved. But mere consciousness just recognizes the present and that it appears to change – it is not until creative intelligence evolves that the past and the future are recognized and the importance of time to evolution is understood. While they have no direct effect on each other, space could not exist without time and time could not exist without space.

But, while space and time are obvious and must be, and require no explanation, matter and energy don't fit in that same category –they don't have to be, and we must provide an explanation for how they came to be. My suspicion is that over infinite and eternal time and space, on average, matter and energy and their products don't actually exist. We are merely a temporary product of random chance, a probability blip that will eventually disappear – unless … (see Chapter 6).

So, while I have argued that time and space must be, indeed can't not be, on the other hand I argue the opposite for matter and energy. They mustn't be, indeed can't be, and so their existence is a mystery. There seems to be no good explanation. The God theory is childish, unsubstantiated and proves nothing (you must ask who or what created God and encounter an endless progression). The concept of gods may have derived from the ancients' experience with extraterrestrials, as some have suggested, but even if that is so, that still doesn't explain how they or we came to be. But there is a possible way to explain the creation of matter and energy acceptable within the scriptures of quantum mechanics and consistent with the concept that on average they don't really exist in the long term. Let's review this possibility.

We know energy appears to be conserved, so there are two possibilities regarding energy:

1. Energy is eternal and fixed (and of course can't be infinite or there would be no need for the law of the conservation of energy).
2. The energy in the universe averages to zero. This would explain the need for the law of the conservation of energy – it is necessary to maintain the average at zero.

I choose the second of these possibilities, particularly because this explains the necessity for the conservation of energy. And, furthermore, it eliminates the issue with option number 1, i.e., why is there energy, and where did this permanent energy come from in the first place? I will come back to this. As for matter, it seems there may be a probability, although very low, of matter being created temporarily from nothing – the very low probability being overcome by the infinity of space and the infinity of time. Indeed, the reality, infinity and eternity of space and time are essential for the emergence (even though temporary) of matter and energy, and eventually us.

Matter could be created as virtual positive/negative pairs of particles (i.e., matter-antimatter pairs) that usually annihilate. This is not a new concept, but generally accepted in quantum physics. While the occasional occurrence of such "virtual" particles wouldn't lead to the creation of the matter universe, the power of the infinity of space and the infinity of time could allow for the unlikely event of an immense number of such pairs occurring simultaneously (a reasonable incorporation of chance/probability into the physics of the universe). This could lead to the creation of the universe we now observe, as I describe next.

Most of these particle/antiparticle pairs would annihilate with the production of energy and photons. This would be the source of the uniform cosmic microwave background (CMBR) noise we observe today. This energy could also facilitate the separation of the remaining particle/antiparticle pairs such that the separate entities would go their own way and gradually clump and create sufficient gravity to begin working against the repulsive positive energy that I will discuss in more detail later. The energy release from the massive and approximately simultaneous particle/antiparticle annihilations could be considered a sort of Big Bang. They may be an even more realistic Big Bang in a sense, because the Big Bang of the Big Bang Theory is of unknown origin, doesn't explain the uniformity of the CMBR and as presented would provide a centre to the universe which apparently does not exist.

I must digress here a bit to discuss the concept of a centre. In this discussion, we are referring to a centre as a beginning – that is, a point that represents the origin of the universe. That is what the status quo believes does not exist. When we speak of the "observable universe", as we will shortly, we are talking about a sphere because the "observable universe" must of necessity be a sphere (I hope I am right in assuming that this is too logically obvious to require an explanation). Now, the point I wish to make here is that the sphere representing the "observable universe" would obviously have a centre, but that centre would have nothing to do with origins or beginnings (again I hope this is too logically obvious to require further explanation). Note further that in our case the centre would be the earth – but there would be multiple overlapping and non-overlapping spheres depending on the observer (e.g., on different planets or in different galaxies, etc), and the radius of the spheres would depend on the speed of light and the level of technological sophistication of the observer.

The theory that I will develop going forward avoids this problem of a centre and explains the uniformity of the CMBR, the arrow of time, the flatness of the universe, the apparent accelerating expansion of the universe, and the existence and nature of "dark energy".

This particle/antiparticle theory of course requires an amount of antimatter equal to our known matter "universe" to exist somewhere – the present standard belief is that somehow there was more matter than antimatter at the beginning, and so when annihilation was complete, there was only matter left. But there is no logical basis for this belief, and it seems very unlikely, so we must ask how these two entities got separated. If they collided they would annihilate – so any clumps that formed would be of either matter or antimatter, and any such clumps of one type that came together with the other type would annihilate. Only those clumps that were sufficiently separated from the opposite type would survive and continue to grow.

Antimatter is believed by most physicists to respond identically to gravity as matter does, but some have proposed that matter and antimatter may repel in some circumstances (17, 18, 19). This has not yet been experimentally confirmed one way or the other. But, if matter and antimatter were repulsive, this would provide an explanation for how matter and antimatter got so completely separated and may allow some new speculation on the source of dark matter and dark energy. Indeed, the very distant galaxies that astronomers have observed moving away from us most rapidly may actually be antimatter galaxies that are moving apart faster since they are being repelled rather than attracted. But, given the more likely mutual attraction, and since I'm sure the space universe is infinite and eternal, and therefore our visible matter-filled "miniverse" is just an incredibly tiny fraction of the total universe, there may be many other miniverses of matter and antimatter out there attracting the galaxies in our miniverse (see Fig. 10).

I must digress once again to make an important point. There will be clamours against the proposal I am about to make. Those who have swallowed the present interpretation of the Theories of Relativity hook, line and sinker would say that if light hasn't reached us from these miniverses, then gravity hasn't either. While Newton's theory assumes the speed of gravity to be infinite, many physicists believe that Relativity proves that the speed of light cannot be exceeded, and that the speed of gravity is the same as the speed of light. There are three possible responses to this:

The first is that Newton was right, i.e., that gravity is instantaneous (the speed of gravity is infinite), so there is no problem. I think this is not completely unlikely and am waiting for someone clever enough to prove it.

Second, the speed of gravity doesn't have to be infinite, but merely significantly greater than the speed of light, to be compatible with my proposal. Physicists have reported the first measurement of the speed of gravity (Kopeikin and Fomalont, 23), claiming it is the same as the speed of light. However, examining their report, I suspect that they didn't actually measure the speed of gravity but rather the speed of light, which brought them the information. Stuart Samuel (31) and Clifford Will (22) came to the same conclusion as I did. And keep in mind that I have in previous chapters shown that the conclusion from the Theory of Relativity, that the speed of light cannot be exceeded, is highly suspect. It has been more recently reported, as I reviewed in Chapter 4, that the detection of gravitational waves has been accomplished and confirms the speed of gravity to be the same as the speed of light. I raised doubts regarding what these amazing experimental measurements actually determined.

And third, even if it turns out that the speed of gravity is the same as the speed of light, there is still an explanation that might eliminate the issue. This is that the masses involved in these external miniverses emitted gravity long before they were able to emit light into a transparent or traversable environment.

Let me re-emphasize the point I made before I digressed regarding the speed of gravity – our visible miniverse, as immense as it may seem, is merely a negligible fraction of an infinite universe. Therefore, there may be a huge number of such miniverses out there beyond our visibility, and probably a mixture of matter-made and antimatter-made. And, since gravity has been presumed to act over infinite distance, they could all be exerting some gravitational effect on our miniverse (even if gravity's speed is less than infinite). I am of course unable at present to put an accurate number on the exact size of these effects, but I can say with confidence that these effects could certainly be sufficiently large to explain "dark energy" when you consider the following calculations (based on Fig. 10).

Using Newton's universal law of gravitation ($F = G \times M_1 \times M_2 / r^2$), we can calculate the force of gravity between our miniverse and a miniverse outside our visibility horizon (Fmm), using the following values gleaned from the scientific literature:

1. M_1 = normal mass of our miniverse = ~ 10^{52} kg
2. M_2 = mass of external miniverse = we assume approx. the same = ~ 10^{52} kg
3. r = distance between the centres of the 2 miniverses = ~ 4×10^{29} m (assume 10^3 x the diameter of our observable miniverse, which is ~ 4×10^{26} m)
4. G = ~ 7×10^{-11} m/sec^2

Then Fmm = $7 \times 10^{-11} \times 10^{52} \times 10^{52} / (4 \times 10^{29})^2 = 4.4 \times 10^{31}$ kilonewtons

And also the force of gravity of an external miniverse on one of our galaxies (Fgm), wherein:

1. M_1 = mass of the galaxy = ~ 4×10^{41} kg
2. M_2 = mass of the external miniverse = ~ 10^{52} kg
3. r = distance between centres of the galaxy and the external miniverse = ~ 4×10^{29} m (assume same as 3 above)

Then Fgm = $7 \times 10^{-11} \times 4 \times 10^{41} \times 10^{52} / (4 \times 10^{29})^2 = 1.75 \times 10^{21}$ kilonewtons

Note: It is apparent that these forces are large, which informs us that this possibility for explaining dark energy is feasible and the external miniverses can be vast distances from our "miniverse". I will assume that we are surrounded by many such miniverses, similarly in all directions around our "miniverse", and we might be able to estimate the probable average distance to the nearest surrounding miniverses from the size of the force in kilonewtons exerted by the "dark energy". With this theory I can also suggest that the dark energy will not be an absolute constant, but it should nevertheless be similar, though not likely identical, in all directions. I should note that while the forces I *calculated above seem large, they would represent fields between miniverses that would be much more compatible with the dark energy field than the values (e.g., vacuum energy density suggested by quantum field theory that is off by a factor of 10^{120}) based on the presently accepted interpretations of relativity and quantum theories (32).*

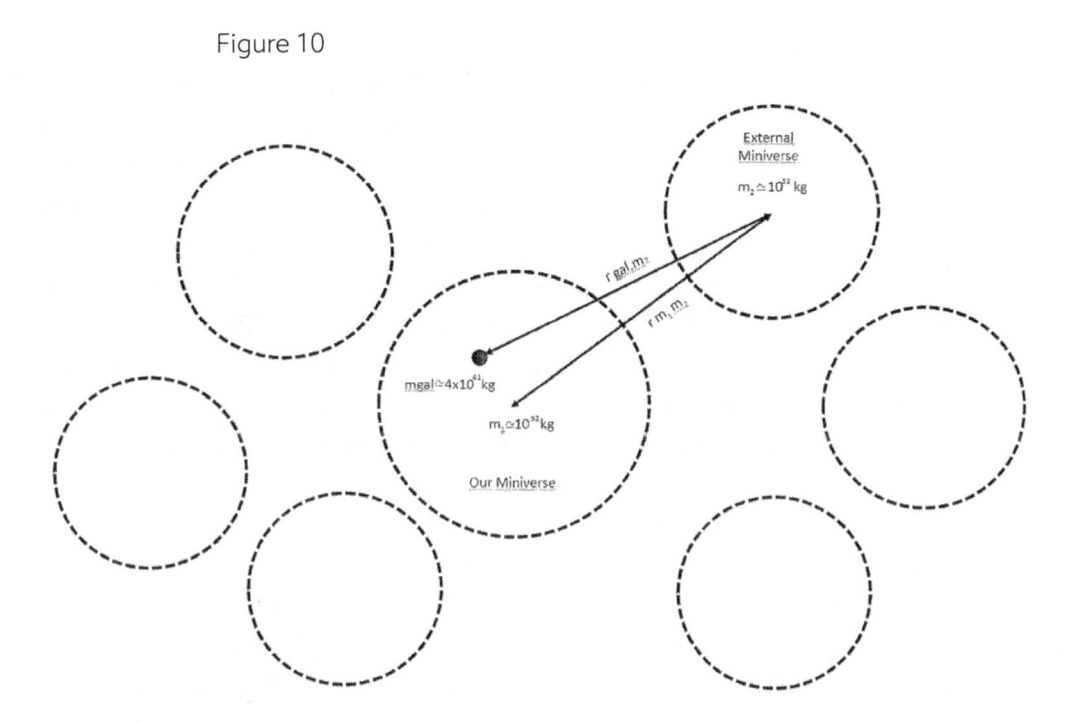

Figure 10

This figure shows the arrangement of the universe that the above calculations, suggesting a possible explanation for "dark energy", are based upon.

Thus, using the estimated values for the mass and the radius of our visible miniverse, and assuming similar values for miniverses beyond our visible horizon, one can estimate the gravitational effects between miniverses by using Newton's equation for universal gravitation. The potential effect is large enough that the distance between the miniverses could be very large without eliminating this effect as a significant factor. And indeed, by using the estimated mass of galaxies we can calculate the gravitational effect of external miniverses on individual galaxies. These effects would be significant and could certainly explain "dark energy".

To repeat, "dark energy" could be, and is most likely to be, merely the result of gravitational effects from miniverses existing beyond the visible horizon of our observable miniverse. And this explanation for the source of dark energy would also explain the fact that dark energy is not constant but changing with time. This would be due to miniverses coming and going as well as the distances (and therefore the gravitational effect) between them, changing as expansion and separation continue. I would suspect that, while the gravitational effect of these miniverses would be roughly uniform in all directions, cosmologists might detect slight variations in different directions. I also suspect that they might eventually be able to estimate the probable number, distance and distribution of these miniverses, even though they are beyond the visual capability of their telescopic technology. Of course, this all might (although not necessarily) depend on the velocity of gravity being significantly greater than the velocity of light.

Here in a nutshell then is what I think explains the existing data, given a better understanding of how relativity and quantum mechanics should be interpreted. Inflation never happened, and the Big Bang didn't occur, at least not as suggested from an infinitesimal point source. In the "beginning" (of matter; there was no beginning of space or time) there was infinite space, and an extremely rare event (but possible within the infinities of space and time) occurred whereby virtual particle/antiparticle pairs were spontaneously produced, which for the most part immediately annihilated with the production of energy and photons that became the CMBR radiation. (Because there was no "Big Bang" from an infinitesimal point, the CMBR radiation is uniform throughout the universe.) Matter, and anti-matter, that escaped annihilation, began to form groupings, then under the influence of the respective gravitational fields gathered to form larger groupings. See note at end of this chapter.

This happened throughout the infinity of the universe and resulted in miniverses distributed throughout the universe, providing an explanation for the observation of "dark energy". Dark energy is described as a relatively weak field that is repulsive, opposite to the effect of the gravitational field. In this new theory, as the groupings of matter (eventually galaxies in each miniverse) move farther apart, the effect of "local" gravity diminishes and the effect of the dark energy (i.e., "distant" gravity from external miniverses mimicking a repulsion in our miniverse) predominates (and indeed increases as the miniverses gradually move closer together). The movement apart of galaxies in our miniverse would therefore accelerate in agreement with what cosmologists observe.

To repeat, it is obvious that time and space are infinite and eternal, i.e., have existed forever, and cannot not be. A similar argument cannot be made for matter and energy. We have just discussed the possibility that matter sprung spontaneously from empty free space as virtual particle/antiparticle pairs. Energy is conserved, implying that it must have existed forever, but this seems unlikely and is impossible to explain. Thus, I favour the suggestion that energy in the universe must average to zero as previously suggested (26). The matter/antimatter scenario may help take care of this necessity. The positive energy provided by matter is balanced by the negative energy of the gravitational field produced by matter. When matter and antimatter eventually annihilate, energy will be annihilated as well. Thus, just as matter exists as equal amounts of matter and antimatter that one day will meet and annihilate, energy may also be a temporary phenomenon that averages to zero and will one day annihilate. Thus, energy in this scenario, like matter, doesn't exist in the long run, and the fact that it must always average to zero may explain the law of the conservation of energy.

I have presented above a theory of how what we see came to be – what we see is not the universe but merely the matter that came to be within our observational capability. I insist that time and space have always been and always will be – indeed must be, cannot not be. Matter, on the other hand, comes and goes and on average might not even exist – so that we are probably just a low probability blip in an ebbing and flowing circus of random chance. Interestingly, when this circus very occasionally produces some relatively short-lived matter, this matter seems to be governed by very strict incontrovertible laws, which probably makes sense given the fact it can't be allowed to wander outside the "boundaries" that maintain it averaging to zero (and similarly for energy).

So, what predictions does this new theory make? Of course this theory is not complete. There are missing parts and I must collaborate with many talented and open-minded scientists to corroborate, change and expand aspects of the theory, along with expert mathematical physicists to see what, if any, new equations or modifications of existing equations might be appropriate, and what unexpected predictions these equations may make. Here, however, are a few "predictions" that the theory is compatible with, without the need for mathematical sophistication:

- The universe is, and must be, flat.
- The CMBR is uniform and relatively homogeneous throughout space.
- The gravity of very large objects in space will bend light.
- There will be no centre of the universe.
- There will be effects on the motion of the galaxies in our miniverse that are not explained by the observable matter (that is, "dark energy" will be observed).
- This "dark energy" will not be an absolute constant over time.
- "Dark energy" will be roughly uniform in all directions, but cosmologists might note slight differences in strength between different directions.
- There will be an apparent expansion of the universe due to an increasing separation of the galaxies.
- The separation of the galaxies (so-called expansion of the universe) will appear to accelerate.
- Dark energy will gradually increase at the expense of dark matter.

Most physicists believe dark matter cannot include antimatter (antimatter, unlike dark matter, would be visible unless existing beyond the visible horizon). If antimatter were eventually shown to repel matter gravitationally, this would definitely exclude antimatter as a contributor to dark matter. I show, nevertheless, in Fig. 11 a scenario in which it could contribute to the effect attributed to dark matter. On the other hand, if the prevailing view that antimatter responds identically as does matter to gravity (i.e., if matter and antimatter attract) is correct, then antimatter would constitute a part of the dark energy possibility I describe (i.e., if there are antimatter miniverses beyond the visible horizon of our matter miniverse, which I believe there must be).

The dark energy field, which I suggest is produced by the miniverses external to our visible miniverse, is claimed to be relatively "weak", but its effect is opposite to the effect of the "local" gravitational field. Therefore, as the groupings of matter (eventually galaxies) move farther apart, the effect of local gravity diminishes and the Dark energy field increases, and the movement apart accelerates. That is, Dark energy becomes more effective as the matter (galaxies in our miniverse) separates and the "local" gravitational effect diminishes. Dark energy is estimated to represent 70% or more of the universe, which really means that it provides about 70% or more of the gravitational effects seen on the galaxies in our miniverse. (If I were a better mathematician, I could probably calculate how the "external miniverses" I suggest produce this 70%.) What constitutes dark matter remains a mystery (although see Fig. 11). However, it will be subject to dark energy just as normal matter is, and therefore will be pulled away from our miniverse – thus dark energy will probably increase at the expense of dark matter.

Figure 11

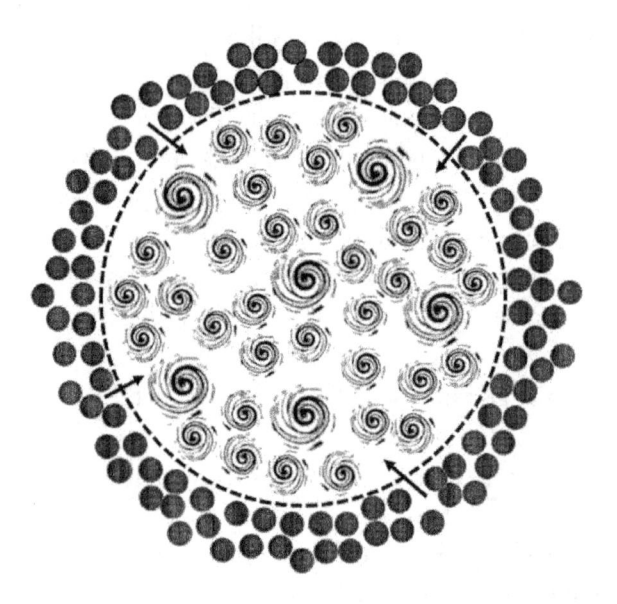

A possible scenario through which antimatter might contribute to "dark matter". If matter and antimatter galaxies got separated during the creation of our miniverse such that antimatter galaxies formed a halo around the matter galaxy core, and antimatter and matter repel, the halo could exert an inward pressure that mimicked or enhanced the gravitational pull in the core. This is very speculative but there is one cosmological observation that could be compatible with this scenario – cosmologists report that the outermost visible galaxies are expanding, i.e., moving apart from the core, faster than nearer galaxies and

that could be explained by the fact that they are being repelled rather than attracted by the inner galaxies. This suggestion is falsifiable – e.g., by confirming that matter and antimatter attract rather than repel.

In reality, this whole somewhat heretical (to some) analysis brings us back to the time when we finally realized that the earth wasn't the centre of the world (what we called the universe back then). There were times when we thought the earth was the centre of everything and everything revolved around "us". Thanks initially to Pythagoras, Copernicus and Galileo, great progress has been made since then, but we still harbour some short-sighted ego that considers our observable/visible miniverse, as THE universe that it is probably not! Based on what I have discussed above, and assuming the reach of gravity complies, it seems possible that our so-called "observable universe" is not necessarily THE UNIVERSE, but merely our visible miniverse, which has a role in the real infinite universe much like a galaxy has in our visible miniverse. It is most likely a negligible speck in the actual infinite universe and it ("we") must be surrounded by many other miniverses as shown in Fig. 10. This might explain the great physics puzzle of our time, "dark energy" (i.e., it is merely the gravitational pull of the surrounding miniverses). This would work based on Newton's universal law of gravitation, but it would be hard to visualize, although possibly more accurately calculated, based on the General Theory of Relativity as it is currently interpreted.

What I am proposing may also explain Einstein's cosmological constant, which physicists have suggested may represent the repulsive force now attributed to "dark energy" but without understanding what this force really is. And the theory I propose suggests that this so-called constant is in fact not a "constant" but will be changing slightly with time, and possibly also with direction.

This new proposal also explains the apparent accelerating expansion of the universe without having to suggest that space itself is expanding. This earlier suggestion was based on the observation that most, if not all, galaxies are moving away from one another, that is moving apart, and moving faster the farther apart they are. The new theory I am proposing explains these observations without having to suggest that space is expanding. By being pulled apart by the gravitational pull of the external miniverses surrounding our miniverse, all galaxies in our miniverse would be moving apart as observed, and this separation would be accelerating. Let me emphasize that, while there seems no reasonable rationale for the suggestion of the expansion of infinite space (infinity cannot be expanded), there is a possible rationale for the separation of matter and energy in this infinite space.

Note, however, that there is a problem with defining the observable limits of our miniverse as the actual boundary of our miniverse. We have no way of knowing if the observable limit is the actual boundary – in fact, it seems highly unlikely that it would occur at that particular point. Fortunately, this thought doesn't change the basic concepts and conclusions of the theory just presented.

But the ultimate outcome from this new theory, compared to that from existing status quo theories, could be very different. Cosmologists presently think that, depending on the rate of expansion, the universe could eventually contract, continue to expand as it is currently doing or accelerate towards a "big rip". The "big rip" depends mainly on the expansion being an expansion of space itself. The "big rip" would not be an outcome of my multiple external miniverse theory as the miniverses would actually be coming together as they pulled their internal galaxies apart. In the presently accepted model of expanding space, there is no explanation for what "force" is causing space to expand, so I prefer my "no big rip" theory.

Let me mention again that there are certainly some gaps and missing parts to the new theory I am suggesting, but the great gains in understanding our universe that we might achieve by pursuing it and filling these gaps I believe justify a serious effort towards this end. The new "no big rip' theory will certainly allow for interesting speculation regarding the beginning and possible future of the universe, as well as of our little miniverse in which our solar system has a limited lifetime. But such speculation will be for a later time when this new theory has been properly vetted.

Note: The Big Bang Theory derives from physicists' acceptance of the present theory of the expanding universe and extrapolating this backwards towards a supposed beginning. My present theory, regarding the cause for the apparent accelerating expansion, doesn't lead to an extrapolation back to such an implausible beginning. This new theory, furthermore, insists rationally that there must be an equal amount of antimatter, as there is matter, distributed throughout the real complete universe. This allows for the possibility that I have proposed that on average, throughout the infinity/eternity of space and time, matter and energy don't really exist and could one day annihilate (unless ... see Chapter 6). Furthermore, a proper understanding and evaluation of the actual distribution of matter and antimatter throughout the real complete universe might eventually provide further explanation for dark matter and dark energy, and the nature and evolution of the real complete universe.

CHAPTER 6

CONCLUDING REMARKS

Here is a succinct summary of what I have attempted to demonstrate in the previous chapters:

1. Mathematics is a powerful tool/language but what it produces is only as good as the premises and data put into it – it is capable of producing misleading results.

2. There is only one universe and it is infinite and eternal in terms of space and time.

3. Space is real, absolute, infinite and eternal, but has no substance or physical properties except our artificial observation and measurement of distance/length. Space is not relative. Only our observation of space (distance/length) is relative.

4. Time is real, absolute, infinite and eternal, but has no substance or physical properties except our artificial measurement of duration. Time is not relative. Only our observation or artificial measurement of time (duration) is relative.

5. Simultaneity occurs and is real, only the observation of simultaneity is relative.

6. Nothing imprinted a direction on time; rather, time imprinted a direction on reality and the evolution of the universe.

7. Matter and energy are not infinite and eternal and may average to zero over infinite and eternal space and time.

8. The Copenhagen Interpretation of quantum mechanics makes no sense and can't be correct, as stated and implied. It refers to the "observed world", not the "real world". The "quantum world" is the "observed world", and unlike the "real world", nothing exists in the "quantum world" until it is observed.

9. Heisenberg's Uncertainty Principle claims that we cannot measure both the position and velocity of a particle beyond a defined level of accuracy, not that a particle does not have a precise position and velocity in reality.

10. The equations of the Theories of Relativity (Special and General) work for certain practical situations but the <u>interpretations</u> of why they work are not correct. When combined with quantum mechanics to explore the origin of the universe, they yield singularities and theories like the Big Bang Theory, Inflation and String Theory, which have serious flaws.

11. The Special Theory of Relativity is not a new law of nature but a slight correction of classical mechanics for certain special circumstances, required because information transfer is not instantaneous, i.e., the velocity of light (carrier of information) is not infinite.

12. Length contraction and time dilation do not actually occur in reality – they are simply errors in observation calculated by the Lorentz Transformation for a given reference frame (and this fact doesn't create preferred frames of reference that would violate any laws or postulates).

13. The General Theory of Relativity is not a correct new law explaining gravity – space cannot be curved and this wouldn't explain gravity even if it were possible. Rather, the ten equations of General Relativity may provide a complicated mathematical correction for Newton's law of universal gravitation by accounting for the observational error arising from changing velocity due to the acceleration caused by gravity. These corrections also account for iterative gravitational fields, which occur because a gravitational field has energy and therefore mass that creates a further gravitational field. Thus, these equations merely calculate, more accurately than does Newton's law of universal gravitation, the curved trajectory in space that will be observed for an object moving in a gravitational field.

14. There are flaws in all the so-called experimental proofs of the General Theory of Relativity. They may show that the equations of the General Theory provide more accurate calculations of phenomena than the simpler single equation of Newton, but do not prove that gravity curves space.

15. There is no "fabric" of spacetime. Spacetime is a derived construct.

16. The prevalent interpretation of Bell's theorem and inequality is problematic. This exercise doesn't really show that reality and locality cannot both exist, but simply demonstrates that quantum mechanics calculates what might be observed. The real world exists without observation, but nothing exists in the quantum (observed) world until an observation is made. If the speed of light were infinite (i.e., if information transfer were instantaneous), then the real world and the observed world would probably be identical and there would be no issue regarding the existence of both reality and locality.

17. Demonstration of all the above has allowed the development of a new theory of the origin and evolution of the universe that accomplishes the following.

    This theory:

    a. Is compatible with what is known:
        i. The universe/space is flat
        ii. The CMBR (Cosmic Microwave Background Radiation) is homogeneous
        iii. There is no centre to the universe
        iv. There is an accelerating separation of galaxies

    b. Explains accelerating separation of galaxies (so-called "accelerating expansion") without requiring the impossible concept that space itself is expanding.

    c. Provides an explanation for dark energy and makes predictions about dark energy that can be tested.

Now, with all this under our belt, let's try to decipher what the rational explanation is for the randomness introduced into physics by quantum mechanics.

It seems likely to be related to the fact that quantum mechanics has been developed to reveal the "observed" world as opposed to the "real" world, as I have suggested throughout this volume. There may be complete determinacy in the real world – but of course this is affected by the uncertainty obviously inherent in any "observed" world (that is, in the observation of the "real" world). The vast majority of our picture of what we call the universe comes from the observation of light and other electromagnetic waves. Light captures a part of the past and keeps it for billions of years. Our remarkable progress in astronomy has shown this. Unfortunately, we are able to recapture only bits and pieces of the total history of events in the universe, and what is recaptured is purely from an observer's point of view. Nevertheless, light does capture, keep and transmit information, possibly forever.

Of course the light hasn't reached us yet from the external miniverses that I have suggested must inhabit the universe outside our visible horizon (and it may not reach earth until long after we are gone), but these external miniverses might provide the "dark energy" that has been puzzling physicists. This of course might, but not necessarily, require that gravity travel faster than light. While many, including Einstein, think that the Special Theory of Relativity proves that nothing can travel faster than light, I have shown that, while the conclusion may be true, it cannot be deduced from the Special Theory of Relativity (i.e., the Lorentz Transformation). Indeed, astronomers have observed that the universe is expanding faster than the speed of light. They attribute this to the expansion of space itself, which they claim is not bound by the limitations on speed provided by relativity theory. But I have explained why space does not and cannot expand – rather galaxies are separating, apparently at rates that exceed the speed of light, and therefore gravity probably does travel faster than light, possibly infinitely fast, and this possibility will be very important to confirm and use.

Observation does not create time, space or matter, but simply gives them meaning. Observation doesn't "collapse waveforms" but pinpoints the reality hidden in the uncertainty of the wave function. There is a reality without observation – otherwise there would be nothing to observe –but observation brings it into consciousness and gives it meaning.

Regarding the birth of the universe: none of the theories advanced so far, for example the Big Bang Theory (see for example, 30), make any rational sense to me. The Big Bang starts with infinite heat and energy confined in an infinitesimal point of space – no explanation is given for this and it is obviously impossible, at least in my estimation. Where did all the heat and energy come from and how can an infinite amount of heat and energy be compressed into an infinitesimal point? We need to develop better theories. I consider it a serious problem when we are told to ignore intuition or common sense, as some proponents of quantum mechanics have suggested. The Big Bang Theory provides no explanation for the creation or origin of the universe it purports to describe – the fact that it can be manipulated to explain a few things that happened after the so-called Big Bang really means nothing, since neither the event itself, nor what caused it or what was before it, is explained (and there are many other things that are left unexplained, such as the flatness of space, the unidirectionality of time, the uniformity of the CMBR, etc.). And Inflation is just as impossible to believe.

We may not really exist in the bigger picture because all matter and energy may average to zero. If on average there is no energy and no matter, there's no need for creation theories because, on average, nothing exists except infinite and eternal space and time, and they must be and can't not be; indeed, they cannot be created or destroyed. We, and what we observe, are just an unlikely blip that will eventually disappear. This may be why the equations of quantum mechanics reflect only probabilities: the universe we observe is merely a probability blip – in the long run it doesn't really exist. Quantum mechanics is not describing reality but merely predicting what might be observed in our ephemeral probability blip.

Since the universe, time and space are all infinite, and must be, and since we know we exist, we can deduce that there have been and will be a possibly infinite progression of civilizations like ours (some inferior, some superior) throughout eternity. So, it seems life can evolve and eventually figure out what is happening, but can probably never change this cycle, and it seems clear that there is no obvious purpose to this endless random cycle that may indeed on average represent nothing (although consider the following paragraph). This may sound depressing, but it really isn't – it gives support to those who simply seek to achieve pleasure and happiness while they can, and there are many of us who achieve these things by struggling to understand what we and our observable environment are all about.

On the other hand, it may be that the creation of matter is natural and must occur, and must progress towards consciousness and increased intelligence and understanding of the natural laws. While it may not be immediately obvious, there may be an unsuspected purpose, a purpose not created by a God, but a natural purpose to eventually create a God. Thus, humankind may be a stepping stone towards the creation of God. What I mean by this is that if we continue to evolve, without destroying ourselves first, we (or at least one of the possibly infinite number of civilizations appearing in the larger universe throughout eternity) may eventually become omniscient, i.e., knowledgeable and creative enough to understand and control the laws of nature. Thus, God did not create man – man is part of the universal process attempting to create a "God". See note at the end of this chapter.

We have so far enjoyed only an incredibly minuscule amount of "universal" time to evolve, and yet we have advanced far enough to be able to observe, think, create and speculate on future dangers and possibilities. Our ability to proceed effectively towards the more appealing possibilities will be affected by a number of things, some under our control, some not yet. Massive earthquakes, massive volcanic eruptions, unexpected asteroid strikes and climate change are among those not presently under our complete control. Surprisingly, given our intellectual and scientific progress so far, one piece not yet adequately under our control is blind belief in unverified religious doctrines and the destructive right-wing conservative mentality. These may take decades or even centuries yet to overcome. Democracy must undergo dramatic evolution to allow such progress – I will discuss this in detail in my next book.

When our civilization eventually disappears, and, on a larger scale, when all civilizations (i.e., life and consciousness) disappear, and when all matter and energy annihilate and cease to exist, infinite space will continue and eternal time will continue – but they will be in a sense meaningless. However, when quantum fluctuations start the process again, time and space will eventually regain meaning. The process will take billions of years to produce consciousness again, but, eventually, that consciousness will give meaning, not only to the time and space in which the consciousness exists, but also to the time between what is thought to be the "beginning" until consciousness finally arose. Of course, if one of the many ephemeral civilizations manages to become "omniscient", then this apparently endless cycle may be altered. Endless cycles are possible in infinite eternities.

Since time is infinite and eternal, there is no starting point and thus no absolute time value. We can only measure time as an interval between two designated points; these points having no meaning on any absolute scale in an infinite, eternal universe. This may have something to do with the fact that time disappears from quantum cosmology, i.e., quantum gravity applied to cosmology. In quantum gravity, quantum mechanics is combined with the equations of General Relativity and falls apart.

A final thought regarding my certainty that the universe must be infinite and eternal: well-trained and experienced physicists may have much more exposure to the theories, mathematics and observations regarding surfaces, manifolds, etc. and be more conditioned to these than I. This allows them to accept things like a finite universe, or a finite unbounded universe, defined by surfaces rather than volumes, 2D objects as being as meaningful as 3D objects and holograms as able to explain apparent reality. Without that conditioning, I see these proposed options as extremely unlikely.

As of this writing, I am 79 years old, and I may have made more mistakes in those 79 years than there are solar systems within 1,000 light years from us, but I am still not much different from that little boy sitting in the closet wondering how and why I was here. If I live long enough, I will write more about the many other things I've wondered about and tried to figure out during these 79 years, which went by so fast that my head is still spinning. Having read (33) about the egos of many very talented physicists, of the infighting for recognition, and knowing how much they have invested in their existing theories and calculations, I expect denigration of what I am proposing. But, of course, I would prefer that these great intellects would instead devote their efforts to help develop and expand these ideas, and even provide equations for them if possible, unless of course they are able to provide convincing and unbiased proof that my theories can't be correct.

I have passed some of this by physicists who have expressed some appreciation for what I am saying but who feel I may be too critical on some aspects. They claim professional physicists may have a better handle on these things than is being expressed in publications for the lay public. This may be so, but there is no justification for publishing nonsense in order to sell books to folks not trained in physics. I know there are brilliant professional physicists out there who realize that there are problems with aspects of the standard interpretation of modern physics (Einstein and Schrödinger were certainly aware of this), and I hope that at the very least what I am proposing will incentivize a few of them to not just criticize or attack what I am saying, but to offer better answers to the issues I raise. I will welcome conclusive proof for or against what I am proposing – but not self-centred offensive or defensive attacks – with magnanimity and gratitude. Indeed, I will find a way to reward any **conclusive** proofs proffered regarding any of the issues I have raised – whether confirming or refuting what I have proposed.

Finally, what I have been saying here has been very succinctly phrased in the following concluding statement from Albert and Galchen's 2009 Scientific American article (34): *"The diminished guru [Einstein] may very well have been wrong just where we thought he was right and right just where we thought he was wrong."* Although Albert and Galchen were proposing something quite different from what I am saying, I am in complete agreement with this clever statement, which I put on the cover of this book.

Final note 1: I need to make the point that probabilities abound even in a so-called deterministic universe. And this plethora of probabilities is essential for the evolution of the components of the universe – especially for the evolution of life, consciousness and intelligence. And the potentially infinite diversity of forms of intelligence throughout the infinity of space and time may eventually result in an "omniscient lifeform" that can fully understand and control natural law and reverse my prediction that matter and energy throughout the universe average to zero – matter and energy may go back to nothing again many times before a "God" is finally evolved.

Final note 2: Unfortunately, in our so-called advanced western democracy, driven too much by excessive capitalism, the desires for self-promotion and enrichment too often overwhelm the effort to learn the truth and benefit the progress of mankind. This unfortunate tendency has reached a crescendo with the recent manifestations of the right-wing conservative Republican element in America. Hopefully the growing progressive liberal Democrat element will begin to reverse this ugly trend and put science at the forefront again.

LIST OF REFERENCES:

1. Scientific American, September, 2015.
    a. The Editors, pp. 32-34
    b. Brian Greene, pp. 34-37
    c. Laurence Krauss, pp. 51-53
    d. Walter Isaacson, pp.38-45
2. Rosenblum and Kuttner, Quantum Enigma, Oxford University Press, 2008
    a. Quote from Werner Heisenberg on p. 104
    b. Quote from authors on p. 104
    c. Quote from John Wheeler on p. 04
    d. Quote from authors on p. 96
    e. Quote from authors on p. 24
3. Neil Turok, *The Universe Within*, House of Anansi Press Inc, Toronto, 2012
4. Brian Greene, *The Fabric of the Cosmos*, Vintage Books, New York, 2005
5. James Gleick, *Chaos: Making a New Science*, Viking, New York, 1987
6. Michio Kaku, *Parallel Worlds*, Anchor Books, New York, 2006
7. Meinard Kuhlmann, Scientific American, August, 2013, pp. 247-255
8. Rafi Moor, *Understanding Special Relativity*, rafimoor.com, 2004
9. Albert Einstein, *Relativity: The Special and the General Theory*, Three Rivers Press, New York, 1961
10. Marcus Chown, *Quantum Theory Cannot Hurt You*, Faber and Faber, London, 2007
11. Peter Cole, "Einstein, Eddington and the 1919 Eclipse", Historical Development of Modern Cosmology, Vol. 252, 2001
12. Paul Halpern, *Einstein's Dice and Schrödinger's Cat*, Basic Books, New York, 2015
13. J. Hecht, "Gravity Probe B Scores "F" in NASA Review", Daily News, May 20, 2008
14. Eugenie Samuel Reich, "Troubled Probe Upholds Einstein", Nature 473:131-132, 2011
15. Paul Dirac, "The Evolution of the Physicist's Picture of Nature", Scientific American, May, 1963
16. P. Marmet and C. Couture, "Relativistic Deflection of Light Near the Sun Using Radio Signals and Visible Light", Physics Essays 12:162-173, 1999
17. Santilli, R.M., "A classical isodual theory of antimatter and its prediction of antigravity", International Journal of Modern Physics 14(14):2205-2238, 1999
18. Villata, M., "CPT symmetry and anti-matter gravity in general relativity", EPL. 94(2), 2011
19. Cobbolet, M.J.T.F., "Elementary process theory: A formal axiomatic system with a potential application as a foundational framework for physics supporting gravitational repulsion of matter and antimatter", Annalen der Physik. 522(10):699-738, 2010

20. Anderton, Roger J., "Einstein's simple derivation of the Lorentz transformation: A critique", The General Science Journal, Oct. 13. 2009
21. Ed Fomalont and Sergei Kopeikin, "The measurement of the light deflection from Jupiter: Experimental results", The Astrophysical Journal. 598(1):704-711. 2003
22. Clifford Will, "Propagation Speed of Gravity and the Relativistic Time Delay", Astrophys. J. 590(2):683-690, 2003 and Stuart Samuel, "On the Speed of Gravity and the v/c corrections to the Shapiro time delay", Phys. Rev. Lett. 90(23):231101
23. Sergei Kopeikin and Ed Fomalont, "Aberration and the Fundamental Speed of Gravity in the Jovian Deflection Experiment, Foundations of Physics 36(8):1244-1285, 2006
24. Rozema et al, Physical Review Letters 109(10):100404 (2012)
25. V. Balakrishnan, Indian Institute of Technology Madras, "Lecture 1: Introduction to Quantum Physics: Heisenberg's Uncertainty Principle", National Programme of Technology Enhanced Learning.
26. Edward P. Tryon, "Is the Universe a Vacuum Fluctuation", Nature 246:396-397, 1973
27. Manjit Kumar, *Quantum: Einstein, Bohr and the Great Debate about the Nature of Reality*, W.W. Norton & Co, N.Y. & London, 2008
28. Relativity notes Shankar, Open Yale Courses, 2006
29. Maury Shivitz, "Logical Derivation of the Lorentz Transformation Equations", drzeesradiorepairblog.biz, 2017
30. Brian May, Patrick Moore and Chris Lintott, "*Bang!: The Complete History of the Universe*", Carlton Books, 2016
31. Stuart Samuel, "On the Speed of Gravity and the v/c Corrections to the Shapiro Time Delay", Phys. Rev. Lett. 90:231101, 2003
32. Hobson, MP, Efstathiou, GP and Lasenby, AN, "*General Relativity: An Introduction for Physicists*", Cambridge Press, UK, 2006
33. Bojowald, M, "*Once Before Time: a Whole Story of the Universe*", Alfred A. Knopf, New York, 2010
34. David Z Albert and Rivka Galchen, "A Quantum Threat to Special Relativity", Scientific American, pp. 32-39, March, 2009
35. Gabor Racz, and Laszlo Dobos, "Explaining the accelerating expansion of the universe without dark energy", Monthly Notices of the Royal Astronomical Society, March 30, 2017
36. J.T. Nielsen, A. Guffanti, S. Sarkar, "Marginal evidence for cosmic acceleration from Type 1a supernovae", Nature Scientific Reports 6, Article # 35596, 2016. See also "The universe is expanding at an accelerating rate – or is it?" Phys. Org (University of Oxford), Oct. 21, 2016.
37. Sabine Hossenfelder, *Lost in Math: How Beauty Leads Physics Astray*, Basic Books, 2018
38. J.G. Von Soldner, "Vorschlagzu eine Grad-Messung in Afrika", Monatliche Correspondenz zur Beförderung der Erdund Jimmelskunde 9:357-62, 1804

39. Michael Brooks, "Wave Goodbye?", New Scientist, pp. 28-32, Nov. 3, 2018
40. Sean Carroll, "Even physicists don't understand quantum mechanics", Opinion, The New York Times, Sept. 7, 2019
41. Albert Einstein – The Essence of Genius. Special Collector's Edition, Harris Publications, 2015
42. Einstein, A., Podolsky, B. and Rosen, N., "Can Quantum-Mechanical Description of Physical Reality be Considered Complete?", Physical Review 47:777-780, 1935
43. Bohr, N., "Can Quantum-Mechanical Description of Physical Reality be Considered Complete?", Physical Review 48:696-702, 1935
44. Bohm, D. and Aharanov, Y., "Discussion of experimental proof for the paradox of Einstein, Rosen and Podolsky", Physical Review 108:1070-1076, 1957
45. Bell, J. S., "On the E-P-R Paradox", Physics 1(3):195-200, 1964
46. Aspect, A., Grangier, P. and Roger, G., "Experimental realization of Einstein-Podolsky-Rosen-Bohm Gedankenexperiment: A new violation of Bell's inequalities", Physical Review Letters 49:91-94, 1982
47. Christian, J. J., "Disproof of Bell's Theorem: Illuminating the Illusion of Entanglement", Brown Walker Press, 2012

APPENDIX 1

PTOLEMY'S THEORY

Claudius Ptolemy lived in Rome sometime in the second century A.D. He was an astronomer, geographer and mathematician who considered the earth to be the centre of the universe. His model of the "universe" was a refinement of previous models developed by Greek astronomers such as Hipparchus. He extended the earlier observations (and the conclusions of the earlier observations) regarding the sun, moon and planets to develop his geocentric theory, popularly known as the Ptolemaic System. He assumed that each planet moved on a small sphere or circle, called an epicycle, that moved on a larger sphere or circle, called a deferent. The stars, it was assumed, moved on a celestial sphere around the outside of the planetary spheres. Ptolemy's major claim to fame was that his model could accurately explain the motions of heavenly bodies. This became the accepted model for understanding the structure of the solar system and his model was used successfully and not seriously challenged for over 1,300 years. But, of course, it was wrong. Much more detail on Ptolemy's model can be found easily online.

APPENDIX 2

SCHRÖDINGER'S CAT

I am lifting a description of Erwin Schrödinger's thought experiment, labelled popularly as "Schrödinger's Cat", from Paul Halpern's excellent book, *Einstein's Dice and Schrödinger's Cat* (12):

"*Schrödinger proposed the thought experiment in 1935 as part of a paper that investigated the ramifications of entanglement in quantum physics. Entanglement (the term was coined by Schrödinger) is when the condition of two or more particles is represented by a single quantum state, such that if something happens to one particle the other(s) are instantly affected.*

"*Inspired in part by a dialogue with Einstein, the conundrum of Schrödinger's cat presses the implications of quantum physics to their very limits by asking us to imagine the fate of a cat becoming entangled with the state of a particle. The cat is placed in a box that contains a radioactive substance, a Geiger counter, and a sealed vial of poison. The box is closed, and the timer is set to precisely the interval at which the substance would have a 50-50 chance of decaying by releasing a particle. The researcher has rigged the apparatus so that if the Geiger counter registers the click of a single decay particle, the vial would be smashed, the poison released, and the cat dispatched. However, if no decay occurs, the cat would be spared.*

"*According to quantum measurement theory, as Schrödinger pointed out, the state of the cat (dead or alive) would be entangled with the state of the Geiger counter's reading (decay or no decay) until the box is opened. Therefore, the cat would be in a zombie-like quantum superposition of deceased and living until the timer went off, the researcher opened the box, and the quantum state of the cat and counter collapsed (distilled itself) into one of the two possibilities.*"

The absurdity of the above suggested implications of the "quantum measurement theory" is obvious. The cat is never in a "zombie-like quantum superposition of deceased and living" except in the minds of the observer and those who would interpret this little exercise as proof of superposition or entanglement at the macro level. And nothing collapsed in reality – only in the minds of those thinking about this. In fact, the cat is alive until the radioactive decay finally occurs, and is dead shortly thereafter. This would be obvious, by intelligent observation as well as logical reasoning, if an interior camera recorded the cat's experience, or a series of "box openings" around the expected decay time observed the cat. Of course this will never be done – to avoid objections thousands of setups would have to be used and multiple openings of one at a time made around the expected time of decay and even then there would be objections. At any rate, this will never be done: if it could be done avoiding all objections, it is obvious what the results would be. Indeed, if numerous such boxes (say 1,000) were opened when the timer went off, the cat would be observed to be alive 50% of the time and dead 50% of the time. Let me repeat: quantum entanglement and collapse occur only in the mind of the observer.

Erwin Schrödinger, who developed the quantum wave equation, and who, like Einstein, detested the absurdities emanating from the Copenhagen Interpretation of quantum mechanics, presented this thought experiment to show the absurdity emanating from this interpretation and the concept of entanglement. However, some have since interpreted and taught it as a demonstration of quantum entanglement embodied even at the macro level. Schrödinger and Einstein continued their allied battle against the "absurdities" until their deaths but never succeeded in developing the preferred "unified theory of everything" that they sought.

APPENDIX 3

ON SPACE AND TIME

It has been repeatedly touted that Einstein's Theories of Relativity ushered in a revolutionary change in the concepts of time and space, and indeed he does disparage the classical concepts of time and space throughout his treatise on *Relativity – The Special and the General Theory* (9). In his section on "The Theorem of the Addition of Velocities Employed in Classical Mechanics", he says *"Let us suppose our old friend the railway carriage to be traveling along the rails with the velocity v, and that a man traverses the length of the carriage in the direction of travel with a velocity w. How quickly, or in other words, with what velocity W does the man advance relative to the embankment during the process? .... in total he covers the distance W = v + w relative to the embankment in the second considered. We shall see later that this result, which expresses the theorem of the addition of velocities employed in classical mechanics, cannot be maintained, in other words, the law that we have just written down does not hold in reality. For the time being, however, we shall assume its correctness."* In fact, it most certainly does hold – it just might not be **observed** correctly from some "frames".

Einstein proceeds further in his section on "The Apparent Incompatibility of the Law of Propagation of Light with the Principle of Relativity" to say: *"There is hardly a simpler law in physics than that according to which light is propagated in empty space. ... c = 300,000 km/sec. ... based on observations of double stars, the Dutch astronomer DeSitter was also able to show that the velocity of the propagation of light cannot depend on the velocity of motion of the body emitting the light."*

He then decides to conduct another thought experiment, but with a ray of light in place of the man he used in the thought experiment described in the first paragraph. He says: *"If a ray of light be sent along the embankment, (the air above it having been removed) we* *see from the above that the tip of the ray will be transmitted with the velocity c relative to the embankment. Now let us suppose that our railway carriage is again traveling along the railway lines with the velocity v, and that its direction is the same as that of the ray of light, but its velocity of course much less. Let us inquire about the velocity of propagation of the ray of light relative to the carriage. It is obvious that we can here apply the consideration of the previous section, <u>since the ray of light plays the part of the man walking along relatively to the carriage</u>.*" NO, THESE ARE NOT THE SAME – the man is being carried along by the carriage – it has exerted a force on him[1] – that's not the case with the ray of light. Thus, in the following calculation, replacing the velocity W of the man relative to the embankment with the velocity of light relative to the embankment, is not appropriate.

Einstein continues: *"The velocity W of the man relative to the embankment is here replaced by the velocity of the light relative to the embankment. w is the required velocity of light with respect to the carriage, and we have:*

w = c − v [i.e., W = v + w = c, or w = c − v]

the velocity of propagation of a ray of light relative to the carriage thus comes out smaller than c. But this result comes into conflict with the principle of relativity." This is a very clever bit of subterfuge by the clever prankster Einstein – I'm surprised so many have fallen for it!

NO – the light is not being carried along by the carriage with velocity v, and therefore we use DeSitter's findings and simply say w = c.

w = W = c, and there is no conflict!

But Einstein continues: *"… it would appear that another law of the propagation of light must necessarily hold with respect to the carriage – a result contradictory to the principle of relativity. In view of this dilemma there appears to be nothing else for it than to abandon either the principle of relativity or the simple law of the propagation of light in vacuo. At this juncture the theory of relativity entered the arena. As a result of an analysis of the physical conceptions of time and space, it became evident that in reality there is not the least incompatibility between the principle of relativity and the law of the propagation of light and that by systematically*

---

1   Some might say – "the man is moving along at constant velocity v so there is no force acting on him". But the train exerted a force on him to get him to the velocity v. The man is being carried along by the train, and, unlike the situation for light, his velocity is affected by his carrier. If he moves at velocity w under his own power while riding on the train his velocity will be v + w relative to absolute space – even though a correction may be necessary to calculate what will be observed from different surrogate frames. On the other hand, the velocity of light relative to absolute space, and all other frames, will be c, because the velocity of light is not dependent on its carrier or emitter (While this is considered proven by DeSitter it is not at all understood how this can be).

holding fast to both these laws a logically rigid theory could be arrived at. This theory has been called the special theory of relativity to distinguish it from the extended theory." This is another clever bit of subterfuge by the clever prankster. The Theory of Relativity is resolving a contradiction that never existed – in fact, what it ultimately does is correct for an error in observation between frames that can occur because information transfer is not instantaneous. Einstein pulled this bit of subterfuge to suggest that he had actually created a new theory or law of nature.

So, Einstein has somewhat misled us here, and continues to do so as he deals with "On the Idea of Time in Physics" (9, p. 25) which suggests that there is no such thing in reality as "simultaneity". I deal with this misrepresentation in Chapter 2 by showing that while Einstein was right in suggesting that simultaneity is **observed** differently from different coordinate systems, it cannot be claimed that simultaneity does not, or cannot, occur in reality.

Einstein proceeds: *"Now before the advent of the theory of relativity it had always been tacitly assumed in physics that the statement of time had an absolute significance, i.e. that it is independent of the state of motion of the body of reference. But we have just seen that this assumption is incompatible with the most natural definition of simultaneity. If we discard this assumption, then the conflict between the law of the propagation of light in vacuo and the principle of relativity disappears."* Again, we are asked to discard a truth in order to resolve a nonexistent contradiction.

THERE IS NO NEED TO DISCARD THIS ASSUMPTION – I SHOWED ABOVE THAT THERE IS IN REALITY NO SUCH CONFLICT. FURTHERMORE, THE ASSERTION THAT TIME IS NOT INDEPENDENT OF THE STATE OF MOTION OF THE BODY OF REFERENCE IS A TERRIBLE MISUNDERSTANDING THAT I AM CONTINUOUSLY TRYING TO CORRECT THROUGHOUT THIS WHOLE TREATISE. The state of motion of a body of reference can have no effect on time – only on the **observation of time** from this body of reference as I show clearly in Chapter 2. This and the also misconstrued new conception of space are two major issues that are causing the terrible problems emanating from quantum mechanics.

Thus, we can keep the assumption that "Time has an absolute significance" – this is necessarily true, but what is also true is that time can't be given an absolute value and there is relativity in the <u>observation</u> of time.

Let me repeat this: "Time has an absolute significance but there is difficulty in giving it an absolute value." This understanding is essential to correcting the incompatibility between the Theories of Relativity and quantum mechanics, and getting past the many other absurdities and singularities that the present interpretation of these theories produce.

Contrary to the accepted dogma since the acceptance of the Theories of Relativity and quantum mechanics, time is real, independent and absolute; it can't be created or destroyed, but rather is infinite and eternal. It seems that this should be intuitively obvious to an open mind not coerced by the mystic dogma emanating from the accepted interpretations of relativity and quantum mechanics. The equations of relativity and quantum mechanics become incompatible when attempts are made to combine them to peer back to the "beginning" or to explain the very small. When they are used to delve into the creation of the universe, they produce singularities and incredible concepts or theories such as the Copenhagen Interpretation, the Big Bang Theory, Inflation, String Theory and multiple parallel universes.

The reason time cannot be assigned an absolute value is that time is eternal and infinite – therefore there is no beginning, no $t = 0$. Time is uniquely unidimensional and unidirectional. It is understandable and usable going forward – we can establish an artificial origin and project forward towards infinity, but looking backwards provides a confusing dilemma. We can look (that is, remember) backwards, but not go backwards. We can't establish where (or should I say when) we are because there is/was no origin, since time is infinite and eternal (and must be). It is impossible to imagine or enumerate where (that is, when) we are, since we are at an infinite distance from the "beginning" of time (indeed there was no beginning).

Again, it is easy to imagine infinity going forward but not so going backward – and this is why physics and cosmology, despite such brilliant progress, are providing such nonsense regarding the "beginning". There was no "beginning of the universe", only a beginning of matter in the universe. The existence of matter allows us to establish surrogate "rigid coordinate systems" to facilitate calculating what we will observe, but not to discern reality. To discern reality and learn where we have come from, we must understand that space and time are real and provide the only absolute benchmarks for motion (which occurs in both space and time). I suggest that we must dispense with the absurd new dogma and learn how to use these reality benchmarks properly.

APPENDIX 4

REALITY OF SIMULTANEITY

In this appendix, I will demonstrate that it is possible to corroborate simultaneity from a moving reference system by means of a simple classical calculation:

If a moving reference body MO (moving with velocity v relative to a stationary origin SO) receives two flashes, the first at time t_1 and the second at time t_2, sent from site A at time t_0' and B at time t_0'' (Fig. 2b), it cannot be determined without more information whether they were emitted simultaneously or not, because there are too many (in fact, an infinite number) of possibilities. With a few pieces of information that narrow down the possibilities, however, it is possible to make that call from MO. For example, if it is known that site A is distance x_1 from SO and that site B is distance x_2 from SO as MO passes SO at time $t_0 = 0$, and MO receives a flash from A at t_1 and a second flash from B at t_2, then the observer on MO could determine whether they were sent simultaneously from A and B, as follows:

101

Figure 2 b

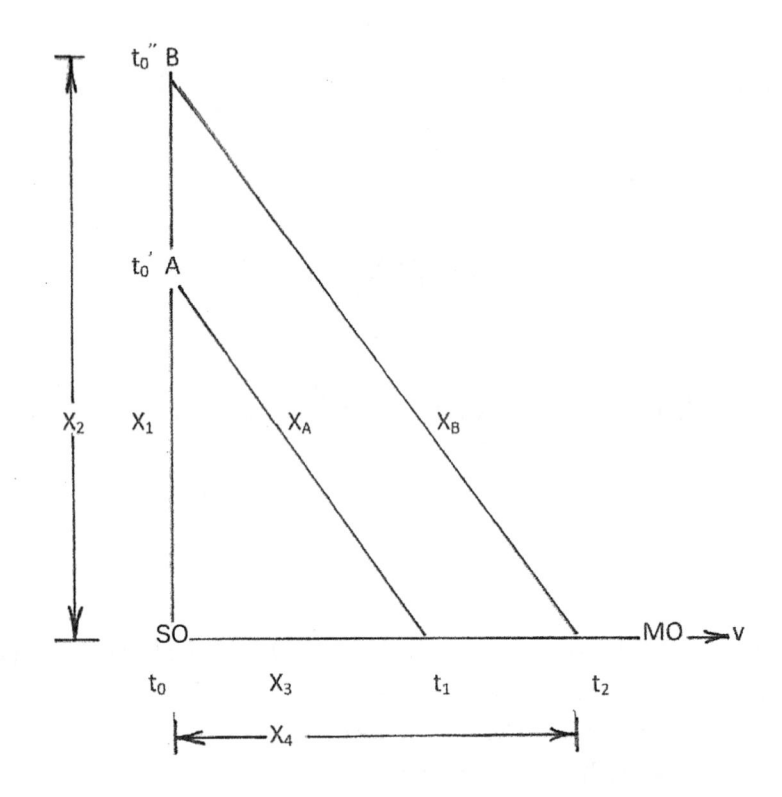

$X_3 = v(t_1 - t_0)$

$X_4 = v(t_2 - t_0)$

$X_A = c(t_1 - t_0')$

$X_B = c(t_2 - t_0'')$

$X_A^2 = X_1^2 + X_3^2$

$X_B^2 = X_2^2 + X_4^2$

Therefore:

and
$$c^2(t_1 - t_0')^2 = X_1^2 + v^2(t_1 - t_0)^2$$
$$c^2(t_2 - t_0'')^2 = X_2^2 + v^2(t_2 - t_0)^2$$

Let $t_0 = 0$, then:

$$(t_1 - t_0')^2 = x_1^2/c^2 + v^2 t_1^2/c^2 = (x_1^2 + v^2 t_1^2)/c^2$$
And $(t_2 - t_0'')^2 = x_2^2/c^2 + v^2 t_2^2/c^2 = (x_2^2 + v^2 t_2^2)/c^2$

Therefore: $t_1 - t_0' = [\text{square root of } (x_1^2 + v^2 t_1^2)]/c$

i.e. $t_0' = t_1 - [\text{square root of } (x_1^2 + v^2 t_1^2)]/c$
and $t_0'' = t_2 - [\text{square root of } (x_2^2 + v^2 t_2^2)]/c$

Thus, t_0' and t_0'' can be calculated to determine if they are equal (all values in the equations are known). Therefore, we have shown, consistent with our intuition, that it is possible to demonstrate simultaneity, with respect to either a stationary or a moving reference body, with the use of nothing but simple logic and mathematics, not incompatible with classical mechanics (or relativity).

APPENDIX 5

NEWTON'S BUCKET

This very interesting and convincing demonstration, which has been the subject of much confusing obfuscation since, was presented by Isaac Newton in 1689. I will start with a presentation of this experiment provided by Brian Greene in *The Fabric of the Cosmos* (4):

"The experiment is this: take a bucket filled with water, hang it by a rope, twist the rope tightly so that it's ready to unwind, and let it go. At first, the bucket starts to spin but the water inside remains fairly stationary; the surface of the stationary water stays nice and flat. As the bucket picks up speed, little by little its motion is communicated to the water by friction, and the water starts to spin too. As it does, the water surface takes on a concave shape, higher at the rim and lower in the center, as in figure 2.1 (and the surface of the water remains concave as long as the water spins, even as the bucket slows and stops)."

The simple and convincing interpretation of this experiment, which Newton realized, is that space is the ultimate arbiter, i.e., the ultimate frame of reference, for motion. It was not the bucket, as seemed likely at first, because the water surface remained concave whether the water was moving relative to the bucket or not. Thus, the motion of the water, causing its reaction to become concave, was relative to space, and space alone. Thus, Newton defined <u>absolute space</u> and claimed an object is truly moving when it is moving relative to absolute space, and an object is truly accelerating when it is accelerating with respect to absolute space.

It is obvious to me that space is absolute, infinite and eternal (must be and can't not be), and apparently Newton himself said, *"Absolute space, in its own nature, without reference to anything external, remains similar and unmovable."* But, although this seems obvious and unassailable to me, apparently many philosophers and physicists throughout the last few centuries have doubted and opposed this fact. Gottfried Wilhelm von Leibniz was one of Newton's most vocal opponents regarding this concept, but Leibniz was eventually forced to admit, *"I grant the difference between absolute true motion of a body and a mere relative change of its situation with respect to another body."*

Greene states, *"This was not a capitulation to Newton's absolute space ..."* I say it was, but Greene has a different take on this! Greene questions Newton's deduction of the existence and role of absolute space and concludes: *"If absolute space really exists, why doesn't it provide a way of identifying where we are located in an absolute sense, one that need not use our position relative to other material objects as a reference point?"* But I don't consider this a reasonable objection: the observation that it doesn't provide a way of identifying where we are located in an absolute sense doesn't prove that it doesn't exist; this is merely the result of the fact that absolute space is infinite and has no defined origin, i.e., there is no $x = 0$, $y = 0$, $z = 0$.

Greene then quotes Ernst Mach's assertions that without material benchmarks for comparison, the very concepts of motion and acceleration cease to have meaning (e.g., you feel acceleration only when you accelerate relative to the average distribution of other material inhabiting the universe). Greene continues by claiming that *"generations of physicists have found it deeply unsettling to imagine that the untouchable, ungraspable, unclutchable fabric of space is really a something – a something substantial enough to provide the ultimate, absolute benchmark for motion. ... space does not enjoy an independent existence."* Yikes, if we put this statement together with the suggestions that "the curvature of space explains gravity" and that "space is expanding and carrying the galaxies along with it" as many claim, one's brain cells start to squirm uncomfortably.

He continues further: *"Only relative motion and relative acceleration had meaning."* But, of course, Mach provided no mechanism or proof for his assertions. Mach, Greene and generations of physicists have continued to argue this point. For my part, I feel certain that space, as Newton demonstrated, is obviously real, absolute, infinite and eternal and furthermore is one of the only two things in the universe that must be, cannot not be and will forever be. The same is true of time. It is in accepting the demotion of space and time where relativity and quantum mechanics have gone off the rails, resulting in singularities and absurdities with respect to reality in spite of the great successes in demonstrating what will be observed in so many situations.

APPENDIX 6

CORRECTION TO FLAWED PHYSICS LESSON REGARDING THE RELATIVITY OF SIMULTANEITY AND TIME

Modified versions of Einstein's train and light-flash thought experiment have been used to teach physics students about the Special Theory of Relativity. In my opinion, these slightly incorrect treatments have helped to promote and preserve the misleading interpretation of the significance of the theory that continues to permeate present-day physics.

I present here a revised treatment of one of these versions that I hope will help promote the proper interpretation of this theory.

Rafi Moor presented this example in his brief paper titled "Understanding Special Relativity" (8), and it has been used to teach physics students about the Theory of Relativity (see Fig. 12):

Figure 12

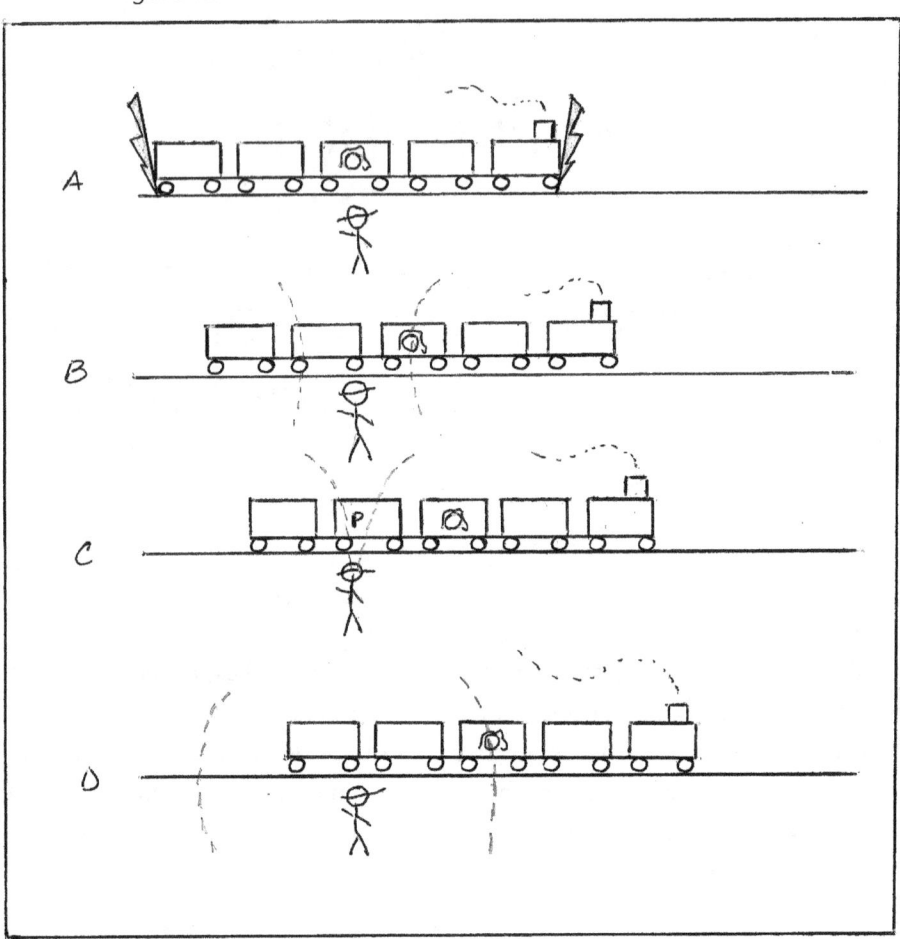

A facsimile of the figure Rafi Moor used to illustrate his version of the Einstein train and embankment thought experiment that I am providing a correction to in this appendix.

"*Suppose there is a train moving at a constant speed along a straight track. A woman is standing exactly in the middle of the train. A man is standing on the ground outside the train. Now, the following scenario is described as seen or measured, or actually happening from the man's point of view:*

"*Just when the woman is in front of him two lightning bolts strike both ends of the train. They leave marks on the train as well as on the track* [this is a seemingly innocuous statement that however misleads Rafi's further analysis of the situation and provides an erroneous lesson to physics students]. *The light from the lightning bolts starts to spread in a constant speed in all directions as shown. A fraction of a second later the light coming from the front of the train reaches the woman as shown. A little later the light coming from both lightning bolts reach the man simultaneously as shown. A little later yet the light from the rear reaches the woman.*" This is correct in reality, but not as perceived from either frame or as Moor continues to present it.

He then states: "*The following phrases are true in the man's frame:*
  a. *The woman is in the middle of the train.*
  b. *The two lightning bolts occur simultaneously.*
  c. *The speed of the light coming from the lightning bolts is the same.*
  d. *The light from the front reaches the woman earlier than the light from the back.*"

Note that while phrase d is true and the woman would have observed this from her "frame", the man would not have noticed this from his "frame". Note further that while Moor claims that these four phrases are true in the man's "frame", he later claims they are not all true in the woman's "frame" – but in fact his claims are not correct. In fact, all four phrases are correct in reality, but the woman would not observe the strikes as simultaneous (as claimed in phrase b) from her frame and the man would not be aware from his frame of the reality stated in phrase d.

He continues: "*Now, let's try to figure out how things happen in the woman's frame. There is one difference we already know: In the man's frame the place where the events of the lightning bolts occur are at the points on the track where they left their marks. In the woman's frame the points are at the edges of the train. Thus, in her own frame, the woman is at all times at the same distance from the points where the lightning bolts hit.*" Yes, where they hit the train, but the marks on the train are no longer the points in space from which the light from the strikes departed towards the woman. This is where Moor goes wrong and arrives at some erroneous conclusions and teaches physics students a very bad lesson.

In fact, the marks on the track do mark the points from which light brings the information regarding the strikes to the man. The man is at all times at the same distance from the points where the bolts hit and from which the light reflected towards him. However, the marks on the train do not serve the same purpose for the woman. The train is moving and the marks on the train are no longer at the actual points in space where the bolts struck. The points on the track mark the actual points in space from which the light brings the information regarding the strikes to the woman (as well as to the man). And since she is moving towards the strike point at the front and away from the strike point at the back, the light from the front bolt will reach her before the light from the bolt at the back (and Moor stated this correctly in the initial descriptive sentences) – the two simultaneous light signals travel different distances to her and so she sees the occurrence of the bolts as not simultaneous. This is because the speed of light is finite and constant in all frames. If the speed of light were infinite, she would have seen the bolts as simultaneous, but the speed of light is not infinite and so the light from the strike at the back of the train would have taken longer to reach her than the light from the strike at the front of the train. Simultaneity can occur in reality, but there is potential relativity in the observation of simultaneity from different frames. The Lorentz Transformation in the Theory of Relativity can correct for this.

The above explanation emphasizes the problem with the dissing of the importance of the benchmark of space. This example provides a convincing way to demonstrate this, but Moor doesn't see or use this opportunity appropriately. In fact, the marks on the track in the "man's frame" provide an appropriate surrogate for the benchmark in space for the events for both observers (those are the points from which the light embarked towards the observers after the strikes), but Moor chooses erroneously to use the marks on the train as the benchmark for the woman's circumstance (but these points have moved along with the woman on the train).

This erroneous choice of surrogate reference point or "frame" may plague quantum mechanics more generally as noted by Lawrence Strauss (1c). He says, "*General relativity tells us that nature is independent of the particular way that scientists choose to define coordinates in space*", then proceeds to say "*many seemingly bizarre results that come out of solving relativity's equations are now understood as mere artefacts of using the wrong coordinate system.*" This seems a serious problem for relativity but what I am discussing here explains this dramatic contradiction. The reality of nature is independent of the way we choose to define coordinates in space, but our **observation** of reality is not.

Moor continues erroneously: "*If we assume that all the phrases above are also true in the woman's frame, there will be a logical contradiction: Light cannot go through equal distances in the same speed and yet not in the same period of time. So, at least one of the four phrases (any of them) must be wrong in the woman's frame.*" No, this is truly misleading – only the fourth phrase (d) might be wrong in reality, but in fact it is not wrong. In fact, the four phrases are all correct in reality and the woman could corroborate them all except phrase b (although phrase b is correct in reality, the woman simply would not have perceived this from her frame).

Based on his false conclusion, Moor proceeds to set up a supposed, and ridiculous, conundrum. He proposes that phrase d is wrong: "*While the man measures the light from the front getting to the woman before the light from the back, the woman sees the light from both sides simultaneously* [**which is of course totally wrong!**]. *This could lead to very strange consequences* [**falsehoods usually do**]: *Suppose we put two photoelectric cells at point P on the train where the two flashes of light meet in the man's frame.* [note: there is a serious problem with this proposal but let's just go with what we assume Moor means here.] *One of the cells is directed to the front of the train and the other to the back. Now we connect the cells to a bomb in a way that if the two cells are illuminated simultaneously the bomb explodes* [correct – but not just in the man's frame]. *In the man's frame the bomb will explode. In the* *woman's it will not since in her frame the flashes meet by her* [wrong – they come at different times – she does not see simultaneity, but this is irrelevant to the issue at hand] *and not at point P.* [**This is totally asinine – the bomb will explode because of the way it was set up. The woman's frame is totally irrelevant to this and if the bomb explodes it will explode in both frames.**] *This will be very hard to settle. … Imagine the man sitting in a bar the next day when the woman enters. 'How come you are here alive?' asks the man surprisingly* [I think he meant surprisedly], *'I saw your train explode to pieces yesterday. There were no survivors.' 'What are you talking about?' says the woman. 'The train got safely to its final station where I got off.'"* **Of course this is complete nonsense** – as I clearly showed the man correctly observed the simultaneity while the woman didn't because of her motion relative to the events. If the bomb had been set as described, she would not have shown up at the man's bar. I am seriously concerned about what kind of physicists this kind of teaching might be turning out. The absurd conundrum Moor presents only appears because of his erroneous reasoning – and the weird situation he describes could never occur. In fact, the impossible outcome he describes confirms his error.

Moor goes on to propose an equally ridiculous scenario that he suggests shows that time is not absolute and independent of reference frame – but this is again total nonsense. The **observation** of time is subject to relativity – time itself is not, time is and must be, absolute and independent.

He goes on to conclude: *"We got now to one of the laws of relativity – the relativity of simultaneity. Two events that are simultaneous in one frame have time difference in another frame."* More correctly, two events that are simultaneous in one frame may not appear to be simultaneous from another frame, the relativity of the **observation** of simultaneity. As I have repeatedly shown, time progresses identically everywhere and in all frames (and must or there would be total chaos in the universe and besides, what could possibly be driving time differently at different points in space?). However, the observation from different frames can differ, depending on certain factors, due solely to the fact that the speed of light is finite and constant in all frames. Simultaneity can and does occur in reality, but, if the circumstances of a frame of observation are such that the distance the photons must travel from each event are different, an error in observation will occur that can be corrected for. The relativity in the observation of simultaneity is due to the lack of instantaneous transfer of information. If the speed of light were infinite there would be no relativity to the observation of simultaneity.

A young physicist friend has suggested an interesting addition to this demonstration, which allows us to strengthen the conclusions I have derived regarding time, space, simultaneity and relativity. In this example (see Fig. 13), the woman resides in the middle of a stationary train car while the man stands on the platform that moves by the train. Just when the woman in the train is directly in front of him two lightning bolts strike both ends of the train. They leave marks on the train as well as on the tracks moving by on the platform. Unlike Moor's analysis of the first (reverse) example, we will attempt to analyze this example correctly from the get-go and see how it relates to, and amplifies, my conclusions from the proper analysis of the first example.

As the bolts strike, the light from each immediately spreads in all directions from the strike points. A fraction of a second later, the light coming from the front of the train reaches the man first (because he is moving towards that point and the light travels a shorter distance). Slightly later the light coming from both strike points reach the woman simultaneously (because she remains equidistant from each strike point). Slightly later yet the light from the rear strike point reaches the man (because he is moving away from this point and this light must travel slightly farther to reach him than did the light from the front strike point).

Figure 13

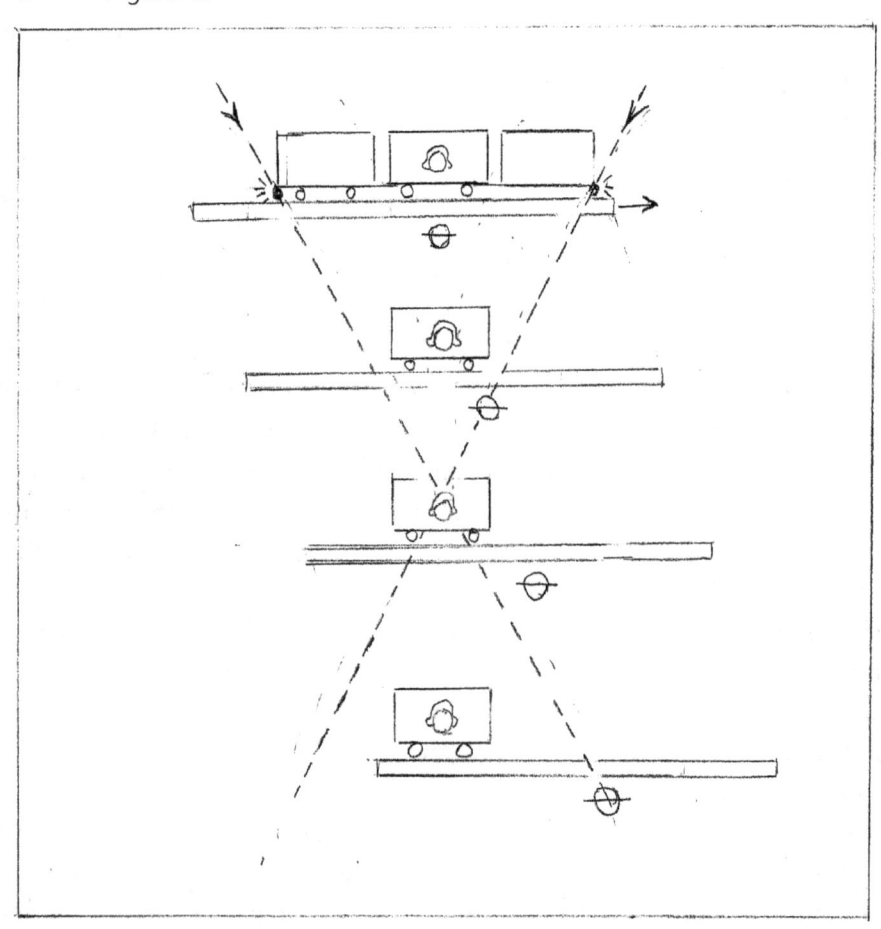

Illustration of my young physicist friend's reverse version of Rafi Moor's version of Einstein's train and embankment thought experiment.

The following phrases are true in the woman's "frame":
- a. The woman is in the middle of the train.
- b. The two lightning strikes occur simultaneously.
- c. The speed of the light coming from both strikes is finite and the same in all directions.
- d. The two simultaneous lightning strikes appear to the woman to occur simultaneously – i.e., the light from both strikes reach her at the same time.
- e. The light from the front strike reaches the man earlier than the light from the rear strike. This phrase is true and the man observes and realizes it, but the woman wouldn't recognize this from her frame. Note that because of the small size of the train and platform relative to the speed of light, the differences we are referring to are minuscule.

The following phrases are true in the man's "frame":
- a. The woman is in the middle of the train.
- b. The two lightning strikes occur simultaneously (although the man does not realize this) and the woman realizes this.
- c. The speed of the light coming from both strikes is finite and the same in all directions.
- d. The two simultaneous lightning strikes do not appear to occur simultaneously to the man – i.e., the light from each strike reaches him at different times.
- e. The light from the front strike reaches the man earlier than does the light from the rear strike. (This phrase is true, and the man observes and realizes it, but the woman does not – she sees the strikes as simultaneous.)

What we notice here is that the results are the reverse of the results in the previous (reverse) example. In the first example the man perceives the simultaneity correctly, while the woman's observation requires correction in order to confirm the simultaneity of the strikes. In the reverse example the woman perceives the simultaneity correctly while the man's observation requires correction. The reason for this is that in the first example the man is equidistant from the two events and is not moving relative to these events, while the woman is moving relative to the events so the two bolts travel different distances to reach her, and the situation is reverse in the second example.

The reason for the observations above is that the speed of light is finite and constant in all frames and directions. In the first example, the light from the two strikes travelled equal distances to the man and therefore appeared simultaneous, but this was not the case for the woman. She was travelling towards the location of the front strike and away from the location of the rear strike, so the light had to travel different distances to her and did not appear simultaneous. The exact reverse happened in the second example.

In point of fact, time is absolute and independent of reference frame, and must be – as I have repeatedly shown. Only the **observation** of time, like the observation of simultaneity, is not absolute or independent of reference frame. Einstein's relativity did not topple "Newton's conception of reality" – nor did it topple the idea that "space and time are absolute and independent". Ours is not a relativistic reality as many have said, but our **observation of reality is relativistic**. The majority of physicists have accepted the erroneous interpretation and claims of relativity for 100 years now and that is why theoretical physics is claimed (1a) to have made no significant progress in all this time, in spite of all the amazing experimental and practical progress that has been made.

APPENDIX 7

ANALYSIS OF THE DERIVATION OF THE LORENTZ TRANSFORMATION PROVIDING INSIGHT INTO THE SPECIAL THEORY OF RELATIVITY

Numerous papers have been published regarding the derivation of the Lorentz Transformation as the basis for Special Relativity, but as stated by Maury Shivitz (29), and many others, *"few of them base their derivation of the Lorentz Transformation equations on a 'logical' footing. ... nearly everyone's derivation of this set of equations is fundamentally a circular argument."* Apparently, Lorentz himself did not derive these equations but found them by trial and error. So, the result of most derivations, including Einstein's, is known before the derivation is done. According to Shivitz, and many others, *"Not only is the argument circular, but peeking at the answer apparently causes almost everyone to overlook a key element of any derivation: it should proceed logically."* I noticed this when I first reviewed Einstein's derivation in his popular book (9). R.J. Anderton (20) had reviewed Einstein's derivation and concluded that: *"Einstein made numerous math mistakes, his special relativity is a collection of math mistakes, and modern physics still uses that collection of mistakes to teach physics students adding extra mistakes as it goes. So, the issue is to try to emphasize one specific mistake among his many which makes it clear that existing SR is a farce needing revision."* I thought this analysis was a bit harsh, particularly since I have such great respect for Einstein, but when I first read Einstein's derivation I was struck by the fact that it seemed to be a bit of purely mathematical hocus pocus, providing little understanding of what it meant. I have provided evidence throughout this book that the Lorentz Transformation, and therefore the Special Theory of Relativity, is merely a correction for the fact that the speed of light is finite and constant in all frames and therefore information transfer is not instantaneous – and this necessitates correction for what is observed from certain frames (i.e., correction to calculations of what will actually be observed from that frame). Essentially there is relativity in the observation of reality, but not in reality itself.

I go one step further in this appendix and explore the derivation of the Lorentz Transformation so as to evaluate my revised interpretation of the Special Theory of Relativity. I do not claim that Einstein's equations are wrong; rather, I wish to analyze further what they actually mean so as to achieve a better evaluation of space and time.

I will start with a slight modification of a version of the derivation provided by Shankar in his Relativity Notes (28). He says: *"an event is something that happens at a definite time and place, like a firecracker going off."* Let us say there is an observer in stationary coordinate system s and an observer in coordinate system s' moving at velocity v to the right compared to s and passing s at time t = t' = 0. Let's say the coordinates of the firecracker blast in s are x, t and calculate what the coordinates for the blast will be in s', i.e., x', t'.

Figure 14

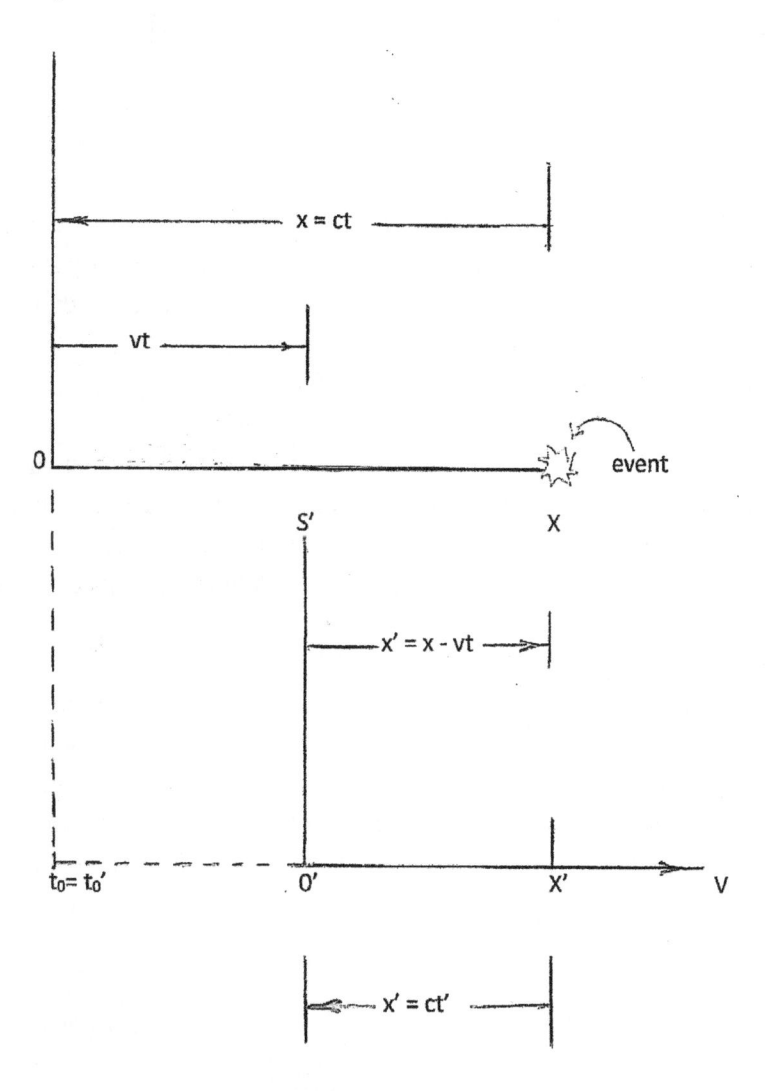

(Figure 14: Diagram illustrating frames S and S' with x = ct, vt, x' = x - vt, x' = ct', showing an event at position x, with axes labeled t₀ = t₀', O', X', V)

Regarding 'logical' derivation of the Lorentz Transformation in the Special Theory of Relativity

Looking at Fig. 14, Shankar suggested that it seemed clear that $x' = x - vt$ (I will show further on that there is a slight problem with Shankar's approach, but for now let's proceed as did Shankar, with one exception – we will determine t' with existing information as well as after calculating lambda).

So now let's determine t' as a function of t.

First we start with the suggested relation: $x' = x - vt$
And since the speed of light is finite and constant in all frames $x = ct$ and $x' = ct'$

117

From these we can easily determine that $ct' = ct - vt = t(c - v)$

i.e., $t' = t(1 - v/c)$

Thus, we have what we might conclude must be the classical transformations for x' and t'. But this is a little different from what Shankar stated. He said: *"First, once we synchronize clocks at t = t' = 0, they remain synchronized for all times t = t'. This is the notion of absolute time we all believe in our daily life."* I believe that there is a general misunderstanding of this notion that I will continue to try and demonstrate. Let me try and clarify at this point that it is dt and dt' that will be always equivalent, but not necessarily t and t' if there is a reason for seeing them differently. I will get to this eventually but let me continue with the standard story Shankar was telling. He proceeds to say the following:

If a. x' = x - vt

and b. x = x' + vt' (note there is something wrong here already since the first equation (a) says x = x' + vt and both this and b can't be right (except in the special case where t' = t, which occurs only when v = 0 or c = infinity) – but Shankar chooses x = x' + vt' without saying how he determined this, and this facilitates the following derivation which wouldn't work using x = x' + vt)

Then, assuming these are correct classical formulas, he asks: *"How are these two equations modified post-Einstein* [implying that space and time are no longer absolute]. *If the velocity of light is to be the same for you and me* [ie, both observers], *it is clear we do not agree on lengths or times or both* [this is not an absolute truth]. *Thus if I predict you will say the event is at x' = x - vt, you will say that my lengths need to be modified by a factor lambda so that the correct answer is*

c. x' = lambda(x – vt)

Likewise, when you predict I will say x = x' + vt' I will say, "No, your lengths are off, so the correct result is

d. x = lambda(x' + vt')

…. *The "fudge factor" for converting your lengths to mine and mine to yours are the same lambda. This comes from the postulate that both observers are equivalent."* This last statement may be correct even though the introduction of lambda is very arbitrary and as I will eventually show not quite appropriate in this circumstance.

Then he calculates lambda by multiplying the left side of these equations and equating this product to the product of multiplication of the right sides giving:

xx' = lambda²(xx' + xvt' – x'vt – v²tt') and using x = ct, x' = ct' he

gets   c²tt' = lambda²(c²tt' – v²tt') which can be solved to get

lambda = 1/square root of (1 – v²/c²)

Now let's calculate t' as did Shankar. Substituting 1/square root $[1 - v^2/c^2]$ for lambda in equations c and d and solving for t' he gets

t' = (t - vx/c^2)/square root of $[1 - v^2/c^2]$

i.e., t' = t(1 - v/c)/square root of $[1 - v^2/c^2]$ since x = ct

But, of course, this result would not have been obtained had the correct equation d been used, i.e., x = lambda(x' + vt) instead of x = lambda(x' + vt')

and notice that for v much less than c, i.e., v^2/c^2 close to 0, this reduces to

t' = t(1 - v/c), identical to what I determined above without relativity considerations and that will approximate t' = t when v is much less than c, which of course is the most common situation, and will occur exactly when v = 0 or c is infinite.

This seems to make sense since s' is travelling towards the light flash and therefore would see the flash before s and therefore t' would be less than t, regardless of relativity considerations. But if v = c then t' becomes 0. This can't be right – in fact, if s' were travelling at the same speed as the light, s' and the light flash would meet in the middle, i.e., t' = t/2. Therefore, there must be further error in the model Shankar is using, as I indicated earlier. Let's continue examining this and try to make sense of it,

Let's start with the fact that the speed of light is finite and constant in all frames. The significance of this is that information transfer is not instantaneous and the error in observation thus caused can be handled identically in all frames. Let's see if we can determine lambda more logically and get a better idea of what it means.

Standard derivations, such as that of Shankar, say that there is a difference of unknown cause and meaning between "frames" that can be corrected by the factor they call lambda, and lambda is calculated using pure mathematical manipulation without any attempt at understanding its meaning (as in the case above) – in fact it is often called the "fudge factor" (29). <u>But it has also been stated that it doesn't matter how it is derived – once obtained it can be applied to any generic "event" (29).</u> This is because the speed of light is finite and constant (i.e., the same) in all frames and the statement is correct except for circumstances I will explain further on.

So let's see if we can determine more logically what that lambda factor is that is required to correct the error due to the lack of instantaneous transfer of information. Let's look at the situation for a moving observer, MO, moving at velocity v perpendicular to the line of sight of a stationary observer, SO, and determine the correction factor between them for observing a flash of light. I have already done this exercise (see Fig. 2, Chapter 2, 2b) and the answer is

$$\lambda = 1/\sqrt{1-v^2/c^2}$$

Interestingly, the same value as Shankar obtained with his (and others') "fudge factor" method. But now we may have a better idea what it means.

And now let's review the apparent error in Shankar's methodology, which seems to have been the gold standard up until now, and present a revised approach that should be more accurate.

Let's start as we did earlier in this Appendix mimicking Shankar's approach. but introduce some corrections as we proceed:

Let us say there is an observer in stationary coordinate system s and an observer in coordinate system s' moving at velocity v to the right relative to s along the x axis and passing s at $t = t' = 0$. They both observe the blast of a firecracker, but not at the same time. Let's say the coordinates of the firecracker blast in s are x, t and calculate what the coordinates for the blast will be in s'.

At this point Shankar suggests that it is easy to understand that $x' = x - vt$ and proceeds to do his calculations. Unfortunately this is not quite right as I will explain shortly, but he proceeds in this manner, making another error along the way.

Here is the issue: we have synchronized $t = t' = 0$ when s' passes s, but we have no way to synchronize the flash, i.e., set it off at the same time $t = t' = 0$. So, to get control of this situation, s (or s') must send a light signal to the right along the x-axis at time $t = t' = 0$ to set off the firecracker when it reaches it at x, t.

The light signal takes time t_1 to reach the firecracker and set it off – then the blast takes time t_2 to reach and be recorded by s. But, of course, t_1 must equal t_2 which must equal t, i.e., $t_1 = t_2 = t$.

So clearly $x = ct$ (the coordinates of the firecracker blast are x, t) but s doesn't receive the blast until 2t and s must realize he/she has to divide the time at which he/she receives the blast by 2 in order to get the correct coordinate t, and then determine x as ct.

Now let's determine the coordinates x', t' for the blast in s':

Of course s' receives the blast at $x' = ct'$, but s' has already moved a distance of vt towards the blast at the time t when the blast is set off.

The flash then takes off towards s' at velocity c while s' moves toward x at velocity v.

Figure 15

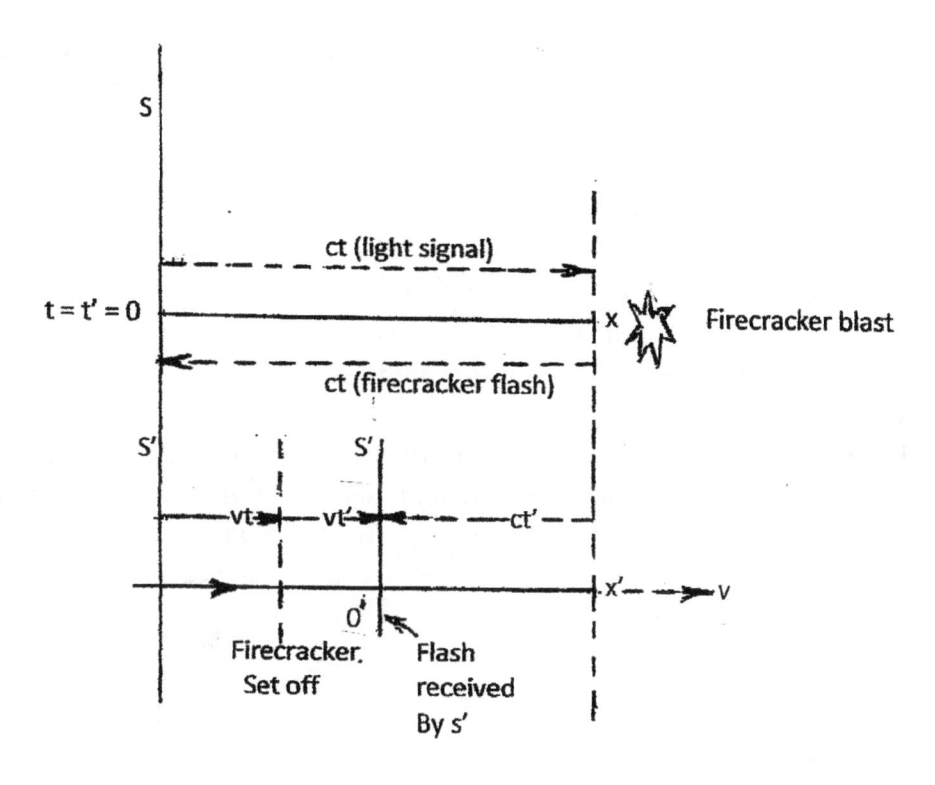

[Figure 15: Revision of Fig. 14 to facilitate determination of coordinates x', t']

Revision of Fig. 14 to facilitate determination of coordinates x', t'

From this new analysis we can see that the following relations hold:
$x' = x - v(t+t')$ $x = ct$ and $x' = ct'$
i.e., $ct' = ct - vt - vt'$
i.e., $t'(c + v) = t(c-v)$
and therefore $t' = t[(c - v)/(c + v)]$

Thus t' is smaller than t as expected and if v = 0, then t' = t as expected

And if c = infinity (i.e., information transfer is instantaneous) then t' = t, again as expected

And furthermore, if v = c, then t' = 0, again as expected because s' would be at x when the firecracker went off. This equation provides the correct calculation of x' and t' for an s' observer moving parallel to the line of sight of the stationary observer, and this holds for velocities v from 0 to c (the velocity of light).

Thus, unlike the original derivation which produced clearly false results this derivation produces correct results and doesn't require lambda.

I will now explain what these results are telling us about the Lorentz Transformation and the Special Theory of Relativity:

The derivation of lambda done in Chapter 2 (c/o Fig.2) in the case of an observer moving at velocity v perpendicular to the line of sight from the stationary observer to the observed object demonstrates the need for the Lorentz Transformation to correct for the fact that information transfer is not instantaneous, i.e., the speed of light c is not infinite, and the light carrying the observation must travel farther to the moving object.

In the case of an observer moving parallel to the line of sight of the stationary observer, as just derived above, the Lorentz Transformation is not necessary, original classical mechanics can make the calculation without the need for the correction.

Of course, in the case of any motion between the parallel and perpendicular to the line of sight of the stationary observer, the Transformation would need to be applied only to the perpendicular arm of the motion vector.

And, again, it is concluded that the Lorentz Transformation and therefore the Special Theory of Relativity is a correction for observations from frames in certain motion relative to the observed object, required because information transfer is not instantaneous, i.e., the speed of light is not infinite. And we do not need to demote absolute space and absolute time as I have continuously demonstrated throughout this book.

APPENDIX 8

EINSTEIN'S FINAL (AND SUCCESSFUL) THOUGHT EXPERIMENT

In this appendix, I will show that Einstein's ingenious last thought experiment, demonstrating the inadequacy of Bohr's version of Quantum mechanics, was in fact successful.

I will reproduce the presentation of this thought experiment from *Quantum: Einstein, Bohr, and the Great Debate About the Nature of Reality* by Manjit Kumar (27, pp 282-288), then present my argument demonstrating that Einstein's point of view was correct, in spite of the unfortunate fact that the physics community has accepted Bohr's dismissal of Einstein's proof.

"*Imagine a box full of light, Einstein asked Bohr. In one of its walls is a hole with a shutter that can be opened and closed by a mechanism connected to a clock inside the box. This clock is synchronized with another in the laboratory. Weigh the box. Set the clock to open the shutter at a certain time for the briefest of moments, but long enough for a single photon to escape. We now know, explained Einstein, precisely the time at which the photon left the box. Bohr listened unconcerned; everything Einstein had proposed appeared straightforward and beyond contention. The uncertainty principle applied only to pairs of complementary variables – position and momentum or energy and time. It did not impose any limit on the degree of accuracy with which any one of the pair could be measured. Just then, with a hint of a smile, Einstein uttered the deadly words: weigh the box again. In a flash, Bohr realized that he and the Copenhagen interpretation were in deep trouble.*

"*To work out how much light had escaped locked up in a single photon, Einstein used a remarkable discovery he had made while still a clerk at the patent office in Bern: energy is* mass and mass is energy. This astonishing spin-off from his work on relativity was captured by Einstein in his simplest and most famous equation: $E = mc^2$, where E is energy, m is mass and c is the speed of light.

"By weighing the box of light before and after the photon escapes, it is easy to work out the difference in mass. Although such a staggeringly small change was impossible to measure using equipment available in 1930, in the realm of the thought experiment it was child's play. Using $E = mc^2$ to convert the quantity of missing mass into an equivalent amount of energy, it was possible to calculate precisely the energy of the escaped photon. The time of the photon's escape was known via the laboratory clock being synchronized with the one inside the light box controlling the shutter. It appeared that Einstein had conceived an experiment capable of measuring simultaneously the energy of the photon and the time of its escape with a degree of accuracy proscribed by Heisenberg's uncertainty principle.

"'It was quite a shock for Bohr,'" recalled the Belgian physicist Leon Rosenfeld, 'He did not see the solution at once.' …. 'During the whole evening he was extremely unhappy, going from one to the other and trying to persuade them that it couldn't be true, that it would be **the end of physics if Einstein were right**,' recalled Rosenfeld, 'but he couldn't produce any refutation.'

"For Einstein it was neither an end nor a beginning. It was nothing more than a demonstration that quantum mechanics was inconsistent and therefore not the closed and complete theory that Bohr claimed. His latest thought experiment was simply an attempt to rescue the kind of physics that aimed to understand an observer-independent reality.

"Bohr spent a sleepless night examining every facet of Einstein's thought experiment. He took the imaginary box of light apart to find the flaw that he hoped existed. … Bohr, desperate to get to grips with the device and the measurements that would have to be made, drew what he called a 'pseudo-realistic' diagram of the experimental set-up to help him.

"Given the need to weigh the light box before the shutter is opened at a pre-set time and after the photon has escaped, Bohr decided to focus on the weighing process. Bohr held that any measurement of the position of the light box would lead to an inherent uncertainty in its momentum, because to read the scale would require it to be illuminated. The very act of measuring its weight would cause an uncontrollable transfer of momentum to the light box because of the exchange of photons between the pointer and the observer causing it to move. The only way to improve the accuracy of the position measurement was to carry out the balancing of the light box, the positioning of the pointer at zero, over a comparatively long time. However, Bohr argued that this would lead to a corresponding uncertainty in the momentum of the box. The more accurately the position of the box was measured, the greater the uncertainty attached to any measurement of its momentum.

"Unlike at Solvay 1927, Einstein was attacking the energy-time uncertainty relation, not the position-momentum incarnation. It was now, in the early hours of the morning, that a tired Bohr suddenly saw the flaw in Einstein's gedankenexperiment. In his desperation to destroy the Copenhagen view of quantum reality, Einstein had forgotten to take into account his own theory of general relativity. He had ignored the effects of gravity on the measurement *of time by the clock inside the light box, gravitational time dilation. The position of the lightbox in the earth's gravitational field is altered by the act of measuring the pointer against the scale. This change in position would alter the rate of the clock and it would no longer be synchronized with the clock in the laboratory, making it impossible to measure as accurately as Einstein presumed the precise time the shutter opened and the photon escaped from the box.*

"*The greater the accuracy in measuring the energy of the photon, via $E = mc^2$, the greater the uncertainty in the position of the light box within the gravitational field. This uncertainty of position prevents, due to gravity's ability to affect the flow of time, the determination of the exact time the shutter opens and the photon escapes. Through this chain of uncertainties Bohr showed* [I would say suggested rather than showed] *that Einstein's light box experiment could not simultaneously measure exactly both the energy of the photon and the time of its escape. Heisenberg's uncertainty principle remained intact, and with it the Copenhagen interpretation of quantum mechanics.*" WRONG!

I will now provide three separate arguments that show that Einstein's experiment was not flawed and therefore the last sentence above is not correct as everyone seems to have assumed.

1. The first argument was presented in Kumar's account: "*Later there would be those who questioned Bohr's refutation because he had treated macroscopic elements such as the pointer, the scale, and the light box as if they were quantum objects and therefore subject to limitations imposed by the uncertainty principle. To handle macroscopic objects in this way ran counter to his insistence that laboratory equipment be treated classically.*" The suggestion that "*Bohr had never been particularly clear about where to draw the line between micro and macro ...*" is too weak to refute the fact that this argument seriously challenges "Bohr's refutation" that Einstein's experiment successfully demonstrated inconsistencies with the Copenhagen Interpretation of quantum mechanics. The next two arguments confirm this.

2. If one merely attached the laboratory clock to the outside of the light box at the same vertical position as the clock inside the light box, the supposed (and horrendously minuscule) desynchronization of the two clocks would be eliminated.

3. Bohr's argument depends on the reality of time dilation and I have shown that time dilation is not a reality. It doesn't really occur; it is merely an error in observation, due to the **relativity of observation**. For the time dilation that is suggested by the Special Theory of Relativity, I showed that this is merely an observational error due to the lack of instantaneous transfer of information due to the finite and constant speed of light. Gravitational time dilation would be a similar observational error, but a little more complicated to calculate because the velocity is not a constant but a changing parameter under acceleration. This argument bolsters

arguments 1 and 2, which make the case even if I am wrong about gravitational time dilation.

Final conclusion: Einstein was right, Bohr et al were wrong – quantum mechanics is not complete and consistent, and the Copenhagen Interpretation may be providing much of the ignored absurdities emanating from quantum mechanics.